SpringerBriefs in Global Understanding

Series editor

Benno Werlen, Department of Geography, University of Jena, Jena, Germany

The *Global Understanding* Book Series is published in the context of the 2016 International Year of Global Understanding. The books in the series seek to stimulate thinking about social, environmental, and political issues from global perspectives. Each of them provides general information and ideas for the purposes of teaching, and scientific research as well as for raising public awareness. In particular, the books focus on the intersection of these issues with questions about everyday life and sustainability in the light of the post-2015 Development Agenda. Special attention is given to the inter-connections between local outcomes in the context of global pressures and constraints. Each volume provides up-to-date summaries of relevant bodies of knowledge and is written by scholars of the highest international reputation.

More information about this series at http://www.springer.com/series/15387

Margaret E. Robertson
Editor

Communicating, Networking: Interacting

The International Year of Global Understanding - IYGU

OPEN

 Springer

Editor
Margaret E. Robertson
School of Education
La Trobe University
Bundoora, VIC
Australia

ISSN 2509-7784　　　　　　　ISSN 2509-7792　(electronic)
SpringerBriefs in Global Understanding
ISBN 978-3-319-45470-2　　　ISBN 978-3-319-45471-9　(eBook)
DOI 10.1007/978-3-319-45471-9

Library of Congress Control Number: 2016949561

© The Editor(s) (if applicable) and The Author(s) 2016. This book is published open access.
Open Access This book is distributed under the terms of the Creative Commons Attribution 4.0 International License (http://creativecommons.org/licenses/by/4.0/), which permits use, duplication, adaptation, distribution and reproduction in any medium or format, as long as you give appropriate credit to the original author(s) and the source, provide a link to the Creative Commons license and indicate if changes were made.
The images or other third party material in this book are included in the work's Creative Commons license, unless indicated otherwise in the credit line; if such material is not included in the work's Creative Commons license and the respective action is not permitted by statutory regulation, users will need to obtain permission from the license holder to duplicate, adapt or reproduce the material.
The use of general descriptive names, registered names, trademarks, service marks, etc. in this publication does not imply, even in the absence of a specific statement, that such names are exempt from the relevant protective laws and regulations and therefore free for general use.
The publisher, the authors and the editors are safe to assume that the advice and information in this book are believed to be true and accurate at the date of publication. Neither the publisher nor the authors or the editors give a warranty, express or implied, with respect to the material contained herein or for any errors or omissions that may have been made.

Printed on acid-free paper

This Springer imprint is published by Springer Nature
The registered company is Springer International Publishing AG
The registered company address is: Gewerbestrasse 11, 6330 Cham, Switzerland

Series Preface

We are all experiencing every day that globalization has brought and is bringing far-flung places and people into ever-closer contact. New kinds of supranational communities are emerging at an accelerating pace. At the same time, these trends do not efface the local. Globalization is also associated with a marked reaffirmation of cities and regions as distinctive forums of human action. All human actions remain in one way or the other regionally and locally contextualized.

Global environmental change research has produced unambiguous scientific insights into earth system processes, yet these are only insufficiently translated into effective policies. In order to improve the science-policy cooperation, we need to deepen our knowledge of sociocultural contexts, to improve social and cultural acceptance of scientific knowledge, and to reach culturally differentiated paths to global sustainability on the basis of encompassing bottom-up action.

The acceleration of globalization is bringing about a new world order. This involves both the integration of natural-human ecosystems and the emergence of an integrated global socioeconomic reality. The IYGU acknowledges that societies and cultures determine the ways we live with and shape our natural environment. The International Year of Global Understanding addresses the ways we live in an increasingly globalized world and the transformation of nature from the perspective of global sustainability-the objective the IYGU wishes to achieve for the sake of future generations.

Initiated by the International Geographical Union (IGU), the 2016 IYGU was jointly proclaimed by the three global umbrella organizations of the natural sciences (ICSU), social sciences (ISSC), and the humanities (CIPSH).

The IYGU is an outreach project with an educational and science orientation whose bottom-up logic complements that of existing UN programs (particularly the UN's Post-2015 Development Agenda and Sustainable Development Goals) and international research programs. It aims to strengthen **transdisciplinarity** across the whole field of scientific, political, and everyday activities.

The IYGU focuses on **three interfaces** seeking to build bridges between the local and the global, the social and the natural, and the everyday and scientific dimensions of the twenty-first century challenges. The IYGU initiative aims to raise awareness of the global embeddedness of everyday life; that is, awareness of the inextricable links between local action and global phenomena. The IYGU hopes to stimulate people to take responsibility for their actions when they consider the challenges of global social and climate changes by taking sustainability into account when making decisions.

This Global Understanding Book Series is one of the many ways in which the IYGU seeks to contribute to tackling these twenty-first century challenges. In line with its three **core elements** of research, education, and information, the IYGU aims to **overcome the established divide** between the natural, social, and human sciences. Natural and social scientific knowledge have to be integrated with non-scientific and non-Western forms of knowledge to develop a global competence framework. In this context, effective solutions based on bottom-up decisions and actions need to complement the existing top-down measures.

The publications in this series embody those goals by crossing traditional divides between different academic disciplines, the academic and non-academic world, and between local practices and global effects.

Each publication is structured around a set of key everyday activities. This brief considers issues around the essential activities of Communicating, Networking and Interacting, as fundamental for survival and will complement the other publications in this series.

Jena, Germany Benno Werlen
May 2016

Preface

An essential part of everyday life is communication. Interacting with our nearby neighbours helps sustain human existence. Exchanging ideas though contacts with other communities can lead to lifestyle improvements and enhanced knowledge about the world.

> Teach thy necessity to reason thus;
> There is no virtue like necessity.
>
> William Shakespeare *Richard II, Act I, sc. iii*

In 2016 we celebrate the life of William Shakespeare who died 400 years ago. His legacy in countries around the world is testament to the power and quality of his vast collection of plays, poetry and other writings. Much of his appeal is the connection in the texts to daily life and themes that are universal. Nevertheless, such recognition is given to few individuals. Whilst village and local communities remain at the core of our lived experience, nowadays technologies contribute to a broadening social horizon for everyone—regardless of personal circumstances. Digital networks and satellite navigation are making communications with distant places possible for people anywhere and anyplace. Location, language and education are diminishing barriers. People and places are connected via affordable networks, quite often using sustainable energy sources such as solar power. Access to mobile phones means previously considered remote locations for human settlement are able to make contact with neighbouring settlements far removed from their home. The global penetration of digital devices means everyday life, present and future, is inextricably linked with information technologies. Reaching into remote villages of the poorest nations, access to global knowledge repositories is becoming the norm. In just 10 years the world figure has risen from 33.9 mobile phones per 100 in 2005 to 96.9 per 100 in 2014 (The World Bank 2016).

The result is that on the ground lived reality locally reflects opportunities and affordances for improving everyday life that are innovative and remarkable. Local government agencies, financial and economic interests, and NGOs are using mobile applications to extend their reach into regions where ground networks are limited and/or remain non-existent. In brief, the twenty-first century is shaping up to be the historic moment when the entire world's people can both receive information and contribute to knowledge making almost instantaneously. The World Bank dash boards suggest progress towards the global development goals of reducing poverty has been made but the challenges are far from over.[1]

Recognizing that agency is the key for success the impact of technological advances provides hope and optimism for all the world's people. Basic needs and services such as health and education are the beginning of what should be the transformative process. As well, productivity is enhanced with greater affordance to global markets and new networks for distribution of goods and services. These are wealth creating for generations, now, and far into the future.

Local knowledge, customs and values, intersecting with leading edge innovation and practices, can, and are, facilitating better education access; improved health outcomes, and lifetime opportunities aimed at transforming human existence. Issues considered in separate chapters of this publication demonstrate a world entwined with digital technologies at every level of communications. Developments bring expert global knowledge to local communities. In turn, local communities have the tools for disseminating knowledge that is actively improving their lives. Ranging from crowd-sourcing using social media to big data sets, the intersections between these extremes are unique.

The chapters that follow highlight how policies for health, education, social and community enterprises work best when they are responsive to local and traditional knowledge bases. They provide a sample of views that are illustrative of the outreach of communications across diverse fields of endeavour and scholarly activity. Importantly, however, it needs to be stressed that this is an open dialogue reflecting the dynamic and fluid context of global communications in an e-networked world. Many more chapters are needed for future publications.

Contributing authors provide information, ideas and challenges in three sections. Part I provides an overview of the three major variables in this discussion. With its focus on natural systems in the chapter titled Our Natural Systems: The Basis of All Human Enterprise reminds us that whilst planet earth is life sustaining and the foundation of human existence, it is also vulnerable to human abuse. Nevertheless, it is through applications of data monitoring and recording technologies that humans are able to measure the impact of their activities and behaviour changes needed. In the second chapter of this trilogy in the chapter titled Technology Trends: Working Life with 'Smart Things' overviews the current e-landscape including trends noted for future innovations. The human connection in the trilogy is considered in the chapter titled Citizenship, Governance and Communication in the form of citizenship, personal identity and what it means to be socially responsible.

Part II includes two examples of how innovative uses of communication technologies are assisting poorer nations face challenges for a better life. Overcoming the human tragedy of the Ebola Crisis in Sierra Leone through health and education

---

[1]See http://data.worldbank.org/mdgs.

is the focus of the chapter titled Changing Cultures: Changing Lives—Mobilising Social Media During a Health Crisis. The chapter titled Bridging the Digital Divide: Everyday Use of Mobile Phones Among Market Sellers in Papua New Guinea considers the agency of knowledge as a means towards enhance agricultural productivity in Papua New Guinea. Taking a different perspective, and representing a slightly outlying angle in this section, in the chapter titled Business, Commerce and the Global Financial System focuses on the mobility of finance and how this reflects economic development. The relevance can be in part explained by the universal measure of development success, or Gross National Product. Monetary measurements provide a standard for successful economic transformation. Connections to development theory highlight the connections between economic advancement and community well-being.

In Part III there are two chapters which examine the power and agency of e-literacy, respectively, for transformative health and education policy. The focus is on Hong Kong and the broader Asian context. However, the issues raised for education are designed to pose questions for communities across the globe. Finally, some recommendations for strategies to broaden and enhance this collection of essays.

Bundoora, VIC, Australia   Margaret E. Robertson
June 2016

Contents

Part I Background and Overview

Our Natural Systems: The Basis of all Human Enterprise 3
Margaret E. Robertson

Technology Trends: Working Life with 'Smart Things' 13
Seng W. Loke

Citizenship, Governance and Communication 21
Michael Williams

Part II Examples of Global Diversity

**Changing Cultures: Changing Lives—Mobilising Social
Media During a Health Crisis** 31
Martha Kamara

**Bridging the Digital Divide: Everyday Use of Mobile Phones
Among Market Sellers in Papua New Guinea**...................... 39
George N. Curry, Elizabeth Dumu and Gina Koczberski

Business, Commerce and the Global Financial System 53
Meg Elkins and Liam J.A. Lenten

Part III Recommendations—Networking the e-Society

**Everyday-ing Health Literacy and the Imperative of Health
Communication: A Critical Agenda**............................... 63
Eric Po keung Tsang and Dennis Lai Hang Hui

Imaging an E-future: Education as a Process Towards Understanding 71
Margaret E. Robertson

Epilogue 77
Margaret E. Robertson

Editor and Contributors

About the Editor

Margaret E. Robertson is Professor of Education in The College of Arts, Social Sciences and Commerce at La Trobe University. Margaret is a member of the Steering Committee of IYGU with interests in youth cultures, pedagogical change and transforming education through innovative uses of technologies. Research interests include cross-cultural analyses of young people's views and visions for the future. She has long contributed to the curriculum and research outputs related to Geographical Education.

Contributors

George N. Curry is Professor of Geography at Curtin University. His research and teaching interests are in sustainability in the broad area of rural development in the developing world. Most of his recent research has been in Papua New Guinea examining sociocultural and economic change associated with the transition to a market economy through the adoption of export cash crops

Elizabeth Dumu is a Ph.D. student at the Curtin University. Her thesis is on the use of mobile phones to improve smallholder livelihoods. She is particularly interested in how mobile phone technology can be utilized by small farmers to access banking services and agricultural extension information.

Meg Elkins is a development economist working in the School of Economics, Finance and Marketing at RMIT University. Her research interests include policy evaluation in developing economies. Further areas of interest are social protection, poverty reduction, corruption, well-being, and education. Meg's current research interests are in the area of applied microeconomics investigating: youth labour market outcomes associated with leisure activities and the economic impact of the arts in Melbourne. Her teaching is in the area of innovation and business design and in 2015 she received a RMIT University teaching excellence award.

Dennis Lai Hang Hui is a Lecturer in the Department of Social Sciences at the Hong Kong Institute of Education. His research interests include public health governance and disaster management.

Martha Kamara is originally from Sierra Leone, West Africa, where she studied for her undergraduate degree at the University of Sierra Leone. She has a Ph.D. from the Australian Catholic University and is currently a Lecturer in the School of Education at La Trobe University, Melbourne. Martha has a total of 30 years extensive teaching experience in intercultural education as a teacher, adult educator and researcher. Her research interests include Indigenous educational leadership, gender and diversity in organizations.

Eric Po Keung Tsang is an Associate Professor at the department of science and environmental studies at the Education University of Hong Kong. His research interest lies in environmental policies and education. In the community he serves as the chairman of Green Power and also a member of the advisory council on the environment HKSAR.

Gina Koczberski is a Senior Research Fellow at Curtin University. Her research interests are concerned with agricultural and social change among smallholder households and rural development in PNG. She is involved in several research projects examining socio-economic and agricultural change in rural PNG with an emphasis on how changing demographic, economic and social circumstances influence household relations of production and strategies of commodity crop production.

Liam J.A. Lenten is a Senior Lecturer at La Trobe University. His Ph.D. (2005) interest in the areas of exchange rate determination models and macroeconomic cycles has more recently concentrated on sports and cultural economics. Liam has held visiting positions at: University of Michigan (US); Massachusetts Institute of Technology (US); University of Otago (NZ); Lancaster University (UK); University of Exeter (UK) and Monash University.

Dr. Seng W. Loke is Reader and Associate Professor in the Department of Computer Science and Information Technology at La Trobe University. He leads the Mobile and Pervasive Computing Research Group and has published over 250 research papers. His research interests include not only technical issues but the impact of technology on life, work and education.

Michael Williams is an Emeritus Professor of Education and former Dean of the Faculty of Education and Health Studies and Head of the Department of Education at Swansea University, UK. He has published widely in the fields of geographical and environmental education, teacher education and school development. He has recently completed the co-editing of the Schooling for Sustainability series published by Springer. His most recent project, completed in 2015, was titled Regeneration, Heritage and Cultural Identity: Perspectives from Canada and Wales.

Part I
Background and Overview

Our Natural Systems: The Basis of all Human Enterprise

Margaret E. Robertson

Abstract Making lifestyle adjustments to benefit the Anthropocene are fundamental for human survival. Whilst the powers of twenty-first century communications systems are celebrated there is a cautionary story needed to set the scene for global advancement. Humans are the custodians of planet earth and dependent on its resources for survival. Networking and communication advances assist the monitoring processes for making possible the survival of the planet and its flora and fauna.

Keywords Nature · Urban living · Memory · Community · Values

Natural Systems as Foundational to Global Understanding

Knowing how planet earth functions is fundamental for all human existence. Natural systems fulfil our basic needs for survival. Whilst this statement may appear self-evident the realization that current generations are growing up in residential spaces increasingly removed from their survival sources is cause for concern. On current estimates more than 60 percent of the world's people live in cities. Their likely experience of everyday life is being surrounded by buildings, transport links and service providers concentrated in local neighbourhoods—themselves constructions of urban living. City life transforms rural landscapes into high rise buildings, rapid transit systems, shopping and entertainment centres, paved surfaces, and ever depleting green spaces. Intricate transport and communication systems enable exchanges of supplies, products and knowledge. 'Community' and belonging are constructions of place which urban residents find or locate in their daily interactions with each other and the systems created.

One outcome for humankind's relationship with natural systems is spatial distancing from the source of supply. Urban access to water provides a simple illustration.

M.E. Robertson (✉)
La Trobe University, Melbourne, VIC, Australia
e-mail: m.robertson@latrobe.edu.au

© The Author(s) 2016
M.E. Robertson (ed.), *Communicating, Networking: Interacting*,
SpringerBriefs in Global Understanding, DOI 10.1007/978-3-319-45471-9_1

Tap and bottled water, for instance, are critical for urban survival. However, their delivery to urban populations via collection points such as dams and reservoirs must flow through often complex network systems, including political borders in addition to purification and recycling plants, and various distribution systems. All of these flows are dependent on communication systems for transfer connections to their final destination points. Put simply, the biosphere landscape is being reengineered to suit our urban demands. Of course history shows that wherever human settlement has taken a grip on the landscape water diversion techniques are evident—for human consumption as well as for stock and crops. Viaducts, simple irrigation systems and water channels are the lifeline of many rural and village communities in more remote locations in present times. They remain as remnants of a lifestyle balance with nature that is largely gone. Today the scale of reordering of nature to meet urban demands, has, and is, reshaping our heritage. Illustrative are the hydro-electric dam projects, such as the Three Gorges Dam across the Yangtze River. The question for all humankind is at what cost for the planet?

For the purposes of this discussion three interacting forces are acknowledged. Each is impacting on the fragility of planet earth, and can, in part, be explained by information systems, flows and networks. Knowledge-based, and reliant on human usage the interoperability of these global systems has irrevocability changed the order of daily living. Predictability and certainty that were once hallmarks of a settled society are now fluid constructs which require community responsibility. Readiness to change is fundamental for learning how to live sustainably in our networked world. In brief the forces are:

- First, the natural ecological balance has been disturbed with resultant climate and atmospheric events, natural disasters including floods, fires, coastal erosion, defoliation and desertification. Making adjustments to benefit the Anthropocene can be considered fundamental for human survival.
- Second, the concentration of people into megacities exacerbates the disturbance of natural systems and adds to air, water and soil pollution.
- Third, the built landscape alters the imaginaries of people. Cities *are* reality. Farming life and food production are increasingly large scale commercial enterprises. The links between people, the land and the landscapes created, reflect affluent hegemonies where daily life has departed from rural subsistence forever. Decision making lenses need to be reminded of our dependence on, and fragile links, with nature.

What We Know About Planet Earth and How Technology Is Helping

Scientists argue about the status of the Anthropocene in geological time (Gibbard and Walker 2013; Rull 2013). The Holocene period marked the beginning of human activity on earth including agricultural development, towns and cities, and migration across continents. Paleontologists study plant and animal remains to determine historic changes in atmospheric conditions. Recent history during the latter part of the twentieth century appears to have entered a period of more marked change (Pawson 2015). This phase we define as the Anthropocene. Ecosystems are being pushed to accommodate massive growth in population, social systems and global networks that penetrate business and commerce; consumption including housing, energy and transport infrastructure; lifestyle choices including education, mobility and family, as well as decisions within macro global and local communities. In just a short period the changes in the earth's systems are remarkable. Some are listed below.

- Climate Change: "Human influence on the climate system is clear, and recent anthropogenic emissions of green-house gases are the highest in history. Recent climate changes have had widespread impacts on human and natural systems."[1]
- Greenhouse gases: The rise of greenhouse gas emission is contributing to increased incidences of drought, fires and tree mortality (Allen et al. 1999; Nurdiana and Risdiyanto 2015; Ren et al. 2015).
- Earthquake activity: Seismic activity monitored by increasingly sophisticated technology is able to forecast changes recorded at local stations. Active Earth Seismology contributes to human understanding of earth tectonics and hazardous locations[2].
- Population increase: "In 1900, world population was 1.6 billion, a total that had taken at least 50,000 years of human history to accumulate. But by the year 2000, world population reached 6.1 billion, in large part because of a dramatic positive achievement: the rapid spread of modern medicine and public health practices after World War II, starting in the 1950s." (World Population Reference Bureau[3])
- Migration and the megacities: The post World War Two Human Rights Declaration has contributed to the movement of people in search of better social, economic and political circumstances. Added to this agreement has been the commodification of transport—particularly air travel. The poverty imbalance appears unaltered with African nations and parts of Asia being amongst the global poorest.

Since life on earth the biosphere and atmosphere have interacted to produce changes. In the past simple measures enabled changes to be recorded. For instance,

---

[1] See Intergovernmental Panel on Climate Change for full report at http://www.ipcc.ch/.
[2] See Active Earth interactive modelling at http://www.iris.edu/hq/programs/education_and_outreach/museum_displays/active_earth/. See also the World Meteorological Organisation at https://www.wmo.int/pages/index_en.html.
[3] World Population Reference Bureau. See http://www.prb.org/Publications/Articles/2011/world-population-7billion.aspx.

Fig. 1 Stevenson's box

climate records relied on manual plotting of temperature, rainfall and pressure data collected from mostly volunteer recorders who used telephone, telegraphic signals and ground transportation to the nearest telephone device. Scattered around the globe the Stevenson's box (see Fig. 1) symbolizes the links between land, communications and human activity. Automated readings from ground fixtures are the major change of recent times. Nowadays global networks share information disseminating from weather stations into websites around the world without interruption.

Technology improvements enable sharing of data globally in real time via satellite networks and wireless enabled portals. Streaming live via Apps to mobile devices anywhere with satellite access can be viewed as distancing the user from the data source and underlying science. Given the end-user experience of smart technologies and personal mobile devices the lived experiences of an ever increasing urban based population could not be more different from those of their ancestral past of less than one hundred years ago. Is the new reality of information readily available 'in the hand' or at the tap of a portable screen good for our earth's future? At one level the answer must be yes. Being informed helps everyone make sound decisions about personal behaviours locally. The converse of this argument is that not knowing about the origins of the data contributes to unreal expectations about planet earth.

Mega Cities and the Millennial Lifestyle

Until the mid-1800s there was relative harmony between humans and planetary systems. The human footprint was manageable. However, the development of machines for manufacturing, cars, trains and aeroplanes, electricity and radio communications marked the start of massive change. In the space of 3–4 generations humans have changed from being motivated by daily survival to consumer addicts. At the heart of much of their behavior is communication networks with incessant temptations to spend, spend and spend more on their personal wellbeing. Capital concentration in cities and the unstoppable modernization of major population nations including China and India is draining finite resources at rates that alarm the earth's scientists. Computer modelling of consumer lifestyles helps inform governments and policy makers of trends that are unsustainable (Peterson and Robertson 2012). Reversing human behavior or finding more sustainable ways to live the millennial lifestyle are major challenges. Data records reveal the stress on resources ranging across our basic needs—water, food and shelter. These are before considering demands for energy to drive electronic devices and development infrastructure including transport, machinery, and basic services including health and education.

Understanding the fragility of these sources of life on earth is a major task for education. Aimed at human survival into the future of the Anthropocene researchers strive to develop efficiencies in production processes that can reduce demand for finite resources. For instance:

- Water footprint: Recycling water in cities and reducing water intake in production processes are remediation strategies. Communicating water limits and education for household efficiencies can help change consumer habits.
- Green energy usage: Government's building policy codes reflect national commitment to the global challenges of climate change and limits on finite resources. Communicating practices through codes, labelling and monitored checks are constructive practices. Government initiatives in many nations include: upgrading electricity to low energy sources, solar paneling, and insulation in buildings, converting utilities to green energy sources, and finding alternatives such as methane gas for low energy usage.
- Public transport infrastructure: Private car ownership is one of the symbols of personal wealth that follows trends in rising GDP. Demand rises thus offsetting clean energy efficiencies. According to the OECD Library: "Since 1990, countries' efforts in introducing cleaner vehicles have been offset by growth in vehicle numbers and the increased scale of their use. This resulted in additional fuel consumption, CO_2 emissions and road building. Road traffic, both freight and passenger, is expected to increase further in a number of OECD countries."[4]

---

[4]See OECD Library at http://www.oecd-ilibrary.org/sites/9789264185715-en/02/03/index.html?itemId=/content/chapter/9789264185715-20-en&mimeType=text/html.

- Renewable energy sources: Energy security is a major issue for modern nations. Challenges include competition for limited resources as well as expanding domestic needs. Non-renewable fossil fuels including coal and gas have increasing demand. However, as nations face complex issues surrounding their emissions and consider the merits of nuclear power sources as a viable alternative, renewables such as wind and solar power are adding substantially to existing renewable systems, including hydroelectricity. The energy grids are enhanced by political, economic and social awareness of the need for ethical and sustainable energy generation. Public awareness of the issues helps drive wise domestic decisions that create hope for the future.

Constructing Better Futures

Global indicators of increasing levels of consumption associated with rising wealth and associated lifestyle expectations mark the reality of the millennial condition. In his May 2015 Encyclical Pope Francis took the significant step of calling on the global community to change their habits:

> We all know that it is not possible to sustain the present level of consumption in developed countries and wealthier sectors of society, where the habit of wasting and discarding has reached unprecedented levels. The exploitation of the planet has already exceeded acceptable limits and we still have not solved the problem of poverty. The mindsets of individual consumers are the 'problem'.[5]

People who currently consume in excess can also provide the solution and fast communication networks can hurry up the process of re-education. Consider, the population 'bomb' scenario described by Paul Ehrlich in the later 1960s could be linked with China's decision to implement its one child per family policy. Communicating and implementing this decision in the large country of China has contributed to the easing of the population bomb potential. In other parts of the world where poverty, high infant mortality rates and low life expectancy rates continue agencies such as the World Health Organization are assisting with family planning and disease eradication programs. Today the impact of population behaviours and associated consumption of global resources can be modelled using informatics tools such as geospatial technologies, mathematical modelling, hydro-modelling, archeological computing, as well as the increasing array of social media communication tools. Whatever the challenges ahead during the Anthropocene period the natural systems and their protection must take precedence. Our survival depends on radical behavior changes such as that which flowed from the Ehrlich (1968) wake-up call regarding population growth.

---

[5]See 'On care for our common home' at http://w2.vatican.va/content/francesco/en/encyclicals/documents/papa-francesco_20150524_enciclica-laudato-si.html.

Half a century later sharing of information is rapid and simple [see Moss et al. (2010) on climate change research]. Ranging from big data to personal communications the advent of satellite communications, underwater cabling and vastly improved service connectivity to the home, plus personal digital devices with tracking agencies, all translates to maximum impact. Disaster management relies on communications to affected locations via social media. The alarm systems of old are relatively inferior in comparison with the message speed via mobile connectivity. Added to this service, information flows related to food and water security as well as health services and agency support are improving the quality of community services globally. Examples of the links between improved communications and management of the planet's natural systems demonstrate effective ways in which humans can minimize their footprint.

Example 1: Urban Agroecology

In a nation where urbanization is a major feature of recent decades, China has been able to regulate and plan for ecological change. Publications within the Chinese scientific academy report on the innovative practices and smart usage of geo-spatial technologies that are reshaping city landscapes. As opportunities for developing new green resources become possible community action can help maintain and expand their usage. In older cities such as Beijing the issues are more problematic (see Chi et al. 2015) and like mega-cities in other global locations layers of the past along with the rapidity of immigration contribute to the urgency for solutions.

Helping to solve the problem, in part, a new green revolution in food producing areas is underway. Enhancing production and yields, Japan, for instance, offers leadership in the new era of sharing knowledge through easily established communication networks. Business economics working in conjunction with agro-science as well as biosecurity measures are revolutionizing production and product distribution.

Example 2: Green Cities; Community Gardens

Along with creative applications of informatics in urban design are innovative strategies for 'making' green spaces. Whist the allotment or small private and community gardens are part of the history of European cities, roof top and vertical gardens are examples of add-on features to buildings in contemporary design. Apartment dwellers who take up gardening related activities are not only becoming more self-reliant but are part of the sustainable urban living solution which includes reducing urban pollutants into the atmosphere.

http://www.theage.com.au/victoria/extreme-gardening-on-the-92nd-floor-of-the-eureka-tower-20150320-1m45sm.html.

Example 3: Building Community

As the process of urban growth and renewal systematically removes future generations from the rural landscapes of their ancestral pasts the matter of community and personal identity are becoming increasingly problematic. Cyclical rhythms of nature are a feature of community practices across cultures. Rituals are often associated with seasonal changes and moon cycles—these events in nature trigger rituals and myths that contribute to the collective memory of the people and their places. Communities evolve with, and from, these shared landscape memories with traditions and shared values expressed through language, art and architecture, habits and preferences. Collective activities may focus around the local park, schools, places of worship, shopping centres and recreation grounds or increasingly through online spaces. However, competing with this coming together for social interaction, more recent neo-liberal market forces of globalization are fueling the commodification of everyday living towards what Foucault (1970) describes as the governmentality of personal agency. The resultant fragmentation of local community building forces in the modern urban settlement can simply just happen with respect for nature being a causality. What we have is an urban ecology in transition from its close connection with nature to technology enabled and sustained built environments. Landscape designers, along with residents and citizens need to imagine new communities, and the affordances of online communications simplify common interest contact. Thirdspace (Oldenburg 1991; Soja 1996) and community networks (Newing 2010) may not have the materiality of past communities. However, they can help maintain the values and virtues associated with living a life well—in harmony with nature, and with respect for fellow beings.

Our Planet

Life on planet earth demands a balance between the natural systems, and human achievements. Digital technologies are perhaps the greatest wonder of our Anthropocene age. They can both model and enable better futures for humanity through relief of poverty and abuse, and/or act as a negative force as the architect of the planet's destruction. Understanding the fragility of planet earth and assisting morally defensible decision making is thus a product of networking and communications.

Open Access This chapter is distributed under the terms of the Creative Commons Attribution 4.0 International License (http://creativecommons.org/licenses/by/4.0/), which permits use, duplication, adaptation, distribution and reproduction in any medium or format, as long as you give appropriate credit to the original author(s) and the source, provide a link to the Creative Commons license and indicate if changes were made.

The images or other third party material in this chapter are included in the work's Creative Commons license, unless indicated otherwise in the credit line; if such material is not included in the work's Creative Commons license and the respective action is not permitted by statutory regulation, users will need to obtain permission from the license holder to duplicate, adapt or reproduce the material.

References

Allen, J., Mssey, D., and Pryke, M. 1999. Eds. *Unsettling cities: movement/settlement*. London: The Open University.
Chi, W., Shi, W. and Kuang, W. 2015. Spatio-temporal characteristics of intra-urban land use change in Beijing, China between 1978 and 2010. *Journal of Geographical Sciences*, 25(1), 3-18.
Ehrlich, P. 1968. *The Population Bomb*. Rivercity Mass.: Rivercity Press.
Foucault, M. 1970. (trans). *The order of things. An archaeology of the human sciences*. New York: Random House..
Gibbard, P.L. and Walker, M.J.C. 2013. The term 'Anthropocene' in the context of formal geological classification. *Geological Society of London, Special Publications*: doi 10.1144/SP395.1
Moss, R., Edmonds, J., Hibbard, K., Manning· M., Rose, S., van Vuuren, D., Carter· T., Emori, S., Kainuma, M., Kram· T., Meehl· G., Mitchell, J., Nakicenovic,· N., Riahi,· K., Smith, S., Stouffer,· R., Thomson, A., Weyant,· J. and Wilbanks, T. 2010. The next generation of scenarios for climate change research and assessment. *Nature*, 463, 747-756.
Newing, H. 2010. Bridging the gap: Interdisciplinarity, biocultural diversity and conservation. In *Nature and culture. Rebuilding lost connections,* eds. S. Pilgrim and J. Pilgrim, 23-40. London: Earthscan.
Nurdiana, A. and Risdiyanto, I. 2015. Indicator determination of forest and land fires vulnerability using Landsat-5 TM data (case study: Jambi Provence). *Procedia Environmental Sciences* 24, 141-151.
Oldenburg, R. 1991. *The Great Good Place*. New York: Marlowe & Company.
Pawson, E. (2015) What sort of geographical education for the Anthropocene? *Geographical Research*, 53(3), 306–312.
Peterson, J., and Robertson, M. 2012. Spatial models as a hub for sustainability education: Exemplifying the transition from producer to user-defined maps in the classroom: In *Schooling and learning for sustainable development An Asia-Pacific regional focus,* ed. M. Robertson, 199-214. Dordrecht: Springer.
Ren, Z., Zheng, H., He, X., Zhang, D.,Yu, X., and, Shen, G. 2015. Spatial estimation of urban forest structures with Landsat TM data and field measurements. Urban Forestry and Urban Greening,14, 336-344.
Rull, V. 2013. A futurist perspective on the Anthropocene. *The Holocene*, 23(8) 1198–1201.
Soja, E. W. 1996. *Thirdspace*. Malden (Mass.): Blackwell.

Technology Trends: Working Life with 'Smart Things'

Seng W. Loke

Abstract This chapter examines current information technology trends, including mobile, wearable and distributed computing, social networks, crowdsourcing, the Internet-of-Things, and social machines, and discuss their current and potential incremental and transformative influence on daily life and work using such technology. We outline several scenarios of working life, and raise questions and issues about the future of the working life.

Keywords Technology trends · Smart things · Ubiquitous work · Crowdsourcing

Introduction

Much has been written about work in the future, the future of work or whether work has a future. End of work stories abound. Oxford University researchers noted that with the development in robotics technology, about 47 percent of U.S. jobs would be at risk, with higher probabilities of computerisation (and so, job losses) in the areas of services, sales and construction occupations.[1] This is a rather alarming figure (for humans) based on the current technological context, considering that robotics technology has yielded products though seems to still have a long way to go before reaching the capabilities in recent movies such as *I, Robot* (http://www.imdb.com/title/tt0343818/) and *Chappie* (http://www.imdb.com/title/tt1823672/).

---

[1]Full report at http://www.oxfordmartin.ox.ac.uk/downloads/academic/The_Future_of_Employment.pdf, September 2013.

S.W. Loke (✉)
La Trobe University, Melbourne, VIC, Australia
e-mail: s.loke@latrobe.edu.au

© The Author(s) 2016
M.E. Robertson (ed.), *Communicating, Networking: Interacting*,
SpringerBriefs in Global Understanding, DOI 10.1007/978-3-319-45471-9_2

This raises questions for countries, such as China, with a much larger labour force and whose economy depends on many labour-intensive jobs. The impact could be larger, though the nature and type of work could shift. Jobs less threatened by machines are those requiring creativity and social skills, and so, creative work and human presence would still be highly valued. What would young people do then, when the jobs are replaced by machines? Perhaps as noted by Martin Ford, the author of the recent book *Rise of the Robots* (Ford 2015), young people in their twenties could get a guaranteed income as a minimal income (possibly by government) and then be encouraged to start businesses to earn on top of that, noting that entrepreneurship would then be encouraged.

As noted in the book *The Second Machine Age: Work, Progress, and Prosperity in a Time of Brilliant* Technologies (Brynjolfsson and McAfee 2016), "in the next twenty-four years", there could be a thousand-fold increase in computer power world-wide, and all humans could be connected via a common digital network, with unprecedented effects on the planet's economics. This invites many questions: for example, what kind of new work or new jobs will such a digital network create? How will people use such a digital network in their work? And will the effects be transformative for all types of work or just some? And if for some, which types of jobs?

In Nagasaki, Japan the Henn-na Hotel is run completely by robots.[2] How far can this idea go? Could we have petrol stations, restaurants, tour groups, and shops completely run by robots? In China, the Internet boom will create 3.5 million employment opportunities by 2020.[3] Hence, while labour-intensive jobs could be replaced by robots in the future, affecting labour-intensive economies, a new digital economy could create new work for the many in the near future.

In the future of work report by PWC,[4] one of three possible scenarios of work in 2022 is the collaboration networks of small organizations—indeed, the power of social networks and similar technologies of the future can facilitate such a scenario. But it is also noted in the report that technology breakthroughs will be the dominant factor that will influence how people work in the future. The rise of Internet based jobs in the digital economy in China as noted above is perhaps evidence in support of this. Future technologies will further transform how we work and what we do, perhaps in some ways unanticipated, simply because some of the technological breakthroughs of the future might be unanticipated.

The notion of loosely collaborative networks of workers fuelled by emerging technology trends seems to be a direction of the future.

---

[2]http://www.abc.net.au/news/2015-08-19/japanese-hotel-run-by-robots/6706822.

[3]This is based on a Boston Consulting Group study, see http://europe.chinadaily.com.cn/business/2015-08/13/content_21583905.htm.

[4]http://www.pwc.com/en/gx/managing-tomorrows-people/future-of-work/assets/pdf/future-of-rork-report-v16-web.pdf.

Changing Trends at Work

There are at least three trends in work worth noting:

- Work will happen anytime anywhere in the future, or at least can happen and will do so if management permits it. The notion of hot-desking, working from home, and working remotely are notions of work that goes beyond the typical scenario of people sitting physically in the same office or building. With advances in virtual reality, high bandwidth communications, wearable devices with various sensors and interaction modes, robotic telepresence, and collaborative systems, the notion of being there, without being there, will become an increasing possibility. People in physical and virtual environments can be co-present, blurring the distinction between the physical presence and the virtual presence.
- Rapid learning on the job will become an increasing possibility. The extreme scenario of getting a job before getting qualifications for it seems absurd. However, many people do have jobs doing work that they learn skills for while on the job. The increasing number of online courses is only a beginning of a shift in learning, when on-demand or on-the-fly learning might be part of work. Having the desired background and related skills will help, but a question remains as to how far rapid learning technologies can go to helping people get equipped on-the-fly. Changes in career pathways accompanied by uptake of new skills will become easier, perhaps only to be slowed down by the need for experience.
- Customising work will be an interesting possibility in the future. A job that needs to be done by someone can be decomposed in a number of different ways. The same job can be done in many different ways and by varying numbers of people, depending on how it is configured. Future technology in coordinating work might accommodate greater flexibility in this matter.

New Work

There are a number of technological trends that could impact work as we know it.

- **Crowdsourcing.** Crowdsourcing is a combination of outsourcing and the crowd, that is, getting work done by a crowd of people, typically via an online platform (Brabham 2013; Howe 2009; Ren et al. 2015), where humans are employed to do computational work that machines might find difficult to do, including translation, image recognition and others. Consider sorting a bunch of pictures of animals according to cuteness—this is not easy for a computer to do but can be done by humans. Another example is sorting a bunch of video clips according to "funnyness". The work on human sorts and joins involves humans in typical database operations (Marcus et al. 2011), and platforms such as Amazon Mechanical Turk[5] allow large jobs to be broken down into small pieces of work (aptly called *microtasks*) to be done by crowds of people. For example, to translate a book, break it up into 100 sections to be translated, each section to translate is packaged as a microtask. Some people, especially, in the developing world, have begun to make a living on such crowdsourcing work alone—they might perform a series of microtasks, each microtask is done in exchange for a small payment. But doing enough of such microtasks and being paid in a strong foreign currency could just be enough to make a living in their world.

With people having idle or spare resources (e.g., a car not so often used or an empty bedroom), and a platform to advertise such resources to the public, and for people to find such resources, new markets might be created (e.g., AirBnB[6] and Uber[7]) (Chase 2015). Indeed, this model can be extended to all sorts of resources, including excess or unused bandwidth, or idle CPUs, on the mobile or at home. People with idle resources can be pooled together to create new markets, enabled by Web and mobile technologies, and people with idle time can work to obtain resources or for monetary benefits. Can this way of marketing idle resources help someone make a living without a 9 am to 5 pm job? Mobile and wearable technologies are allowing such crowdsourcing anytime anywhere but also enabling the context of workers to be an advantage when performing certain work. Simply being at the right place at the right time could be enough qualification for a worker to do a crowdsourced task (e.g., to see if there are parking spots nearby). Indeed, maps of various location-dependent situations can be created, e.g., carpark maps, bandwidth maps, maps of quiet places in a city and so on, via such crowdsourcing and perhaps updated real-time. Such maps could be useful for people to optimise their life, e.g., find carpark spaces faster, find a quiet spot to work, and go to the currently highest bandwidth hotspot, and hence, possibly bring economic benefits to society as a whole.

Social networks will play an important role in crowdsourcing opportunities. An individual belongs to multiple social networks, and while time will tell how they will grow, given the relatively young large-scale social networks that exist today, crowdsourcing tasks and obtaining advice via social networks can become a game-changing approach to problem-solving (Zoref 2015).

Many questions arise as to where crowdsourcing could go. For example, how will this way of making a living scale to richer nations? And what will crowdsourcing enable that previous ways of working could not? Also, how can

---

[5]https://www.mturk.com.
[6]https://www.airbnb.com.
[7]https://www.uber.com.

we measure the economic value of crowdsourced maps of carparks, bandwidth, noise and pollution, and if such maps bring economic benefits to society as a whole, how will contributors be encouraged and compensated? Will the economic value of such maps be high enough to motivate government subsidies (e.g., tax deductions) or payments that will at least initiate the creation of such systems? What new models of markets and work will emerge with such crowd-based models? Can one work full-time sustainably as a contributor, information provider, and helper in his/her own social network (say if the social network has a large enough number of members)? Indeed, crowd work has a future (Kittur et al. 2013).

- **Social Machines, Human and Machine Synergies.** In recent workshops,[8] a paper introducing the notion of social machines (Buregio et al. 2013), and in *Reinventing Discovery: The New Era of Networked Science* (Nielsen 2011), the notion of systems that utilise human-machine synergy has been proposed. Information systems such as Wikipedia,[9] Galaxy Zoo,[10] and EyeWire[11] employ advanced machine processing but also human input in order to scale and to deal with problems current computer algorithms cannot do well in. Where such systems create information that can be reused over and over again, the effort of one person is multiplied many times, yielding a cornucopia of the commons (Loke 2015), as long as individuals are compensated and motivated (perhaps altruistically) to contribute such information. Even when individuals are self-interested, their own contributions contribute towards information bases that eventually benefit themselves as much as it benefits others.

  Questions arise as to how such systems will evolve and develop. For ex-ample, can one work sustainably as part of one or more of such systems, being a contributor to one or more such social machines, and make a living doing that? Also, how will human-robot synergies enable new ways of working and living?

- **Making Smart Things at Home.** Personal fabrication[12] and mass customization have been concepts that current modern technology has made possible. With 3D printers widely available at reasonable costs, and a huge range of printable materials, from plastic, fabric, conductive ink, to biological tissue, there is a large range of highly personalized and customized products that one can make today at home, compared to years ago. There are still relatively high costs of raw materials to be fed into such printers, but the potential for new ways of work this facilitates is greater than ever.

---

[8]See UbiComp 2015 Workshop on Towards Wisdom Computing: Harmonious Collaboration between People and Machines at http://www.irc.atr.jp/en/event/1452/ and social machines at http://www.sociam.org.
[9]Encyclopaedia done by the crowd, https://www.wikipedia.org.
[10]A system for understanding galaxies, http://www.galaxyzoo.org.
[11]A system to map out the neurons in the brain, http://eyewire.org.
[12]See http://www.media.mit.edu/personalfab/ and http://fablabadelaide.org.au/what-is-a-fab-lab\.

There have been various cottage industries, from making cuckoo clocks to artistic wooden products, in settlements in rural Europe to villages in developing countries, but personal fabrication devices such as appropriate 3D printers can amplify creativity and enable new making that wasn't previously possible.

The Internet-of-Things[13] refers to an Internet consisting of things or everyday objects with computational and networking capabilities. Everyday objects, from sprinklers to umbrellas, can be Internet connected and have behaviours adapted to and enhanced with current information. The books *Smart Things: Ubiquitous Computing User Experience Design* (Kuniavsky 2010) and *Enchanted Objects: Design, Human Desire, and the Internet of Things* (Rose 2015) provide an extensive review of the range of products with not only Internet capabilities but also sensors and reasoning capabilities, yielding intelligent objects or smart things that can work together. Producing such smart things or enchanted objects at home will be an interesting scenario of work—a teenage kid could create a smart walking stick (endowed with sensors to capture surrounding information and to provide audio weather reports) for his/her visually-impaired grandfather, and perhaps for all the elderly in his/her village. A mother could design and print out a new digital fabric bracelet for her young daughter, that is not only decorative and comfortable to wear, but doubles as a communication device with basic Internet phone capabilities, using her nephew's electronic 3D component printer and her sister's 3D fabric printer. A range of 3D printers and objects created using 3D printing is gradually emerging in the market place.[14] What new enchanted objects and smart things will be fabricated in new cottage industries?

- **Helping Information Grow.** In the book *Why Information Grows* (Hidalgo 2015), MIT Researcher Hidalgo pointed out the idea that knowledge and know-how is somehow embodied in the social networks of humans. The information worker is one who creates new capabilities, partly by creating new networks of capabilities, "stored" in social networks. The complexity of producing complex products can be aided by appropriate networks of resources that can provide expertise and crowdsourced components. Will future workers contribute to information growth and economic opportunities via such networks? How far can such social networks empower the individual in creating and consuming new information? How can new social networks of capabilities be formed on demand and ad hoc for producing particular products? How does that affect the flow of expertise and the nature of work? What new jobs in the 22nd century will emerge for making information grow?

---

[13]See http://www.theinternetofthings.eu, and http://www.cisco.com/web/solutions/trends/iot/overview.html.

[14]For example, see http://www.cubify.com.

Conclusion

Predicting the future of work is difficult. It does not seem the end of work but perhaps the end of work as we know it; this chapter has only painted a small part of the possible future landscape of work. Despite various ways that work might be forecasted to end, there is new work that could emerge in the future.

A domestic worker could invent and create a new generation of cleaning and tidying-up tools (or robots) or start a business making smart crafts. We have already seen virtual reality glasses made from cardboard[15]; what new innovations within the constraints of an economy can be created by millions of equipped people in developing countries or villages? Future cheap 3D printers which can make use of raw materials available in the natural settings around and within households could empower them. Crowdsourcing and social machines can empower communities, young and old, locally or across international borders, to coordinate and share resources, and to invent tools and solutions, in order to solve local problems or improve efficiencies, in urban environments, but also in villages, from issues of safety as people walk through quiet places at night or walk through long distances of un-policed areas (e.g., to wells to get water), local health issues, to issues of food security and farming. It would not just be harnessing idle human creativity, but creating work by bridging the gap between collective human creativity and everyday life problems. What would happen if schools of bright students (across disciplines) and "unemployed" people in a rich nation are allowed to invent solutions to attack problems and issues in a less educated village in a poor nation? What if such solutions then become customisable packaged solutions forming a basis for a business? Could people with idle time in rich nations be synergised to form the engine of a crowd machine acting as "teachers" mediating education for a village in a poor nation or for poor people in a rich nation? How could technologies amplify the resources of the rich and educated minds so that poor nations and the uneducated will benefit? Could an appropriate knowledge-network platform be built so that villages can exchange knowledge they gained?

It can be argued that technology is neutral but making it work best for all of us is a key challenge for global understanding.

Open Access This chapter is distributed under the terms of the Creative Commons Attribution 4.0 International License (http://creativecommons.org/licenses/by/4.0/), which permits use, duplication, adaptation, distribution and reproduction in any medium or format, as long as you give appropriate credit to the original author(s) and the source, provide a link to the Creative Commons license and indicate if changes were made.

---

[15]https://www.google.com/get/cardboard/, see also 10 gadgets made from cardboard—http://www.pcmag.com/article2/0,2817,2340487,00.asp.

The images or other third party material in this chapter are included in the work's Creative Commons license, unless indicated otherwise in the credit line; if such material is not included in the work's Creative Commons license and the respective action is not permitted by statutory regulation, users will need to obtain permission from the license holder to duplicate, adapt or reproduce the material.

References

Brabham.D.C. 2013. *Crowdsourcing*. Cambridge MA: The MIT Press.
Brynjolfsson, E. and McAfee, A. 2016. *The Second Machine Age: Work, Progress, and Prosperity in a Time of Brilliant Technologies*. New York: W.W. Norton & Company.
Buregio, V., Meira, S. and Rosa, N.2013 Social machines: A unified paradigm to describe social web-oriented systems. In *Proceedings of the 22Nd International Conference on World Wide Web*, WWW '13 Companion, pages 885–890, Republic and Canton of Geneva, Switzerland, 2013. International World Wide Web Conferences Steering Committee.
Chase, R. 2015. *Peers Inc: How People and Platforms Are Inventing the Collaborative Economy and Reinventing Capitalism*. New York: Public Affairs.
Ford, M. 2015. *Rise of the Robots: Technology and the Threat of a Jobless Future*. New York: Basic Books.
Hidalgo, C. 2015. *Why Information Grows*. New York: Basic Books.
Howe, J., 2009. *Crowdsourcing: Why the Power of the Crowd Is Driving the Future of Business*. New York: New Crown Business.
Ju Ren, Yaoxue Zhang, Kuan Zhang, and Xuemin Shen. 2015. Exploiting mo- bile crowdsourcing for pervasive cloud services: challenges and solutions. *Communications Magazine, IEEE*, 53(3):98–105.
Kittur, A., Nickerson, J.V., Bernstein, M., Gerber, E., Shaw, A., Zimmerman, J., Lease, M., and Horton, J. 2013. The future of crowd work. In *Proceedings of the 2013 conference on Computer supported cooperative work* (CSCW '13). ACM, New York, NY, USA, 1301-1318. DOI=http://dx.doi.org/10.1145/2441776.2441923.
Kuniavsky, M. 2010. *Smart Things: Ubiquitous Computing User Experience Design*. Morgan Kaufmann.
Loke., S.W. 2015. On crowdsourcing information maps: Cornucopia of the com- mons for the city. In *Adjunct Proceedings of Ubicomp 2015 (Presented at the 1st International Workshop on Smart Cities: People, Technology and Data)*.
Marcus, A., Wu, E., Karger, D., Madden, S., and Miller. 2011. Human-powered sorts and joins. *Proc. VLDB Endow*, 5(1):13–24.
Nielsen, M.A. 2011. *Reinventing Discovery: The New Era of Networked Science*. Princeton NJ: Princeton University Press.
Rose, D. 2015. *Enchanted Objects: Design, Human Desire, and the Internet of Things*. Scribner, USA; Reprint edition Rose, D. 2015. *Enchanted Objects: Design, Human Desire, and the Internet of Things*. New York: Scribner, Reprint edition.
Zoref, L. 2015. *Mindsharing: The Art of Crowdsourcing Everything*. New York: Portfolio.

Citizenship, Governance and Communication

Michael Williams

Abstract Information and communication technologies (ICTs) are transforming the engagement of citizens in political life at local, regional, national and international levels. Citizenship can be framed in nationally bounded constitutional and legal terms. It can also be framed in a discourse that is not bounded by national boundaries; ICTs facilitate communication between persons and groups who share a common language and sets of concerns. Gender equality, terrorist activity and statelessness are three global issues that highlight changing citizenship issues in the global village and the information-rich global society.

Keywords Agency · Defining citizenship

E-Citizenship and E-Governance

In recent decades, for politicians in some advanced democracies participatory citizenship is associated less with communal action, seen in the supportive work of multiple charities, voluntary agencies and informal social groups, and more with political engagement, especially with voting behaviour. In some countries there has been a marked decline in voters exercising their hard earned right to vote in local, national, and in the European Union, international elections. ICTs are increasingly being seen as an essential way of bridging the divide between the governors and the governed. Of course, the potential value of ICTs in this role is seen differently from either side of the divide. Selected publications, one from Canada and the other from an international organization, illustrate this.

M. Williams (✉)
Swansea University, Swansea, UK
e-mail: michaeltwilliams8@btinternet.com

© The Author(s) 2016
M.E. Robertson (ed.), *Communicating, Networking: Interacting*,
SpringerBriefs in Global Understanding, DOI 10.1007/978-3-319-45471-9_3

- The Canadian publication (Peters and Abud 2009) explores the democratic deficit identified in three trends: declining voter turnout, falling rates of participation in political parties, and declining trust in political leadership. These trends are not unique to Canada. They had been highlighted earlier, for example, in a British government consultation paper (Office of the e-Envoy 2002). Peters and Abud seek a "new and meaningful form of democracy" (p. 8) in which citizens engage in "deliberation and informed participation" (p. 8). Through case studies they show how through e-consultation central government can engage in processes that are alternatives to simple top-down information delivery. Their focus is on civic literacy and the involvement of citizens via ICTs in processes that go beyond consultation to include deliberation and evaluation. While this might be seen as a challenge to governments to respond to rapid changes in the information society there are more profound challenges that accompany the availability of ICTs. On the one hand, technology itself lies in the hands of powerful multi-national corporations, who may or may not be subjected to national controls. On the other hand, messages carried by ICTs may come from anywhere on the globe, and censoring these by governments becomes increasingly difficult. The power of the message has been passing to the people and the power of the technology has passed to private corporations. Through this, the local becomes the global and the global becomes the local. This, of course has a flip side. ICTs carry not only positive messages. They carry propaganda and they have enormous potential for legal and illegal, public and clandestine, surveillance of persons and organizations as they go about their legitimate and illegitimate local, national and international everyday private and public lives.
- The second publication (Economic Commission for Africa 2008) focuses on Africa. While the Canadian discussion has special resonance in the advanced and wealthier nations of the First World, a different context is to be found in the less-advanced and impoverished countries of the Third World. Half a century ago the division between the First and Third Worlds was defined in terms of wealth: financial poverty bringing with it an array of problems, not least in nutrition, health, housing and education. To the infrastructural inadequacies of these countries must now be added the inaccessibility of ICTs. The verbs "Have" and "Have not" must be redefined to take this into account. The authors of an African report (Economic Commission for Africa 1996) warn of the widening gap between information rich and information poor countries with the danger of Africans becoming second class status in a new world order. The ambitious vision of the African Information Society Initiative was of Africa becoming by 2010 "an information society in which every man, woman, child, village, public and private sector office has secure access to knowledge through the use of computers and communication media" (Economic Commission for Africa 2008, p. 8).

Looking Back

The opening sentences in the Publisher's Note to the Pelican version (1944) of Mackinder's *Democratic Ideals and Reality: A Study in the Politics of Reconstruction* read, "In the 20th century we must see things in the big. Statesmen must think in continents, industrialists in world markets". At that time many citizens were either fighting in World War II or seeking simply to survive, often in the most dangerous and difficult of circumstances. Citizens were thinking small, concerned about themselves, their families and their local communities. In the twenty-first century, it could be argued, superficially little appears to have changed, apart from the increasing use of the word globalisation to embrace a wide range of phenomena that are rapidly changing with the introduction of more and more sophisticated information and communication technologies (ICTs). More profoundly, the relationships between individuals and governments, at all levels, have changed and are changing, partly as a response to the dissemination of ICTs. An examination of the concept of citizenship helps to clarify this.

Citizenship Defined

Injecting the word citizenship into any discussion may be compared to throwing a pebble into a pond and watching the ripples spread from the point of impact to the periphery. The ripples reduce in intensity with distance as they enlarge and spread. Citizenship needs to be spatially contextualised, not least because it is most commonly defined in nationally-specific legal terms: citizenship is integral to statehood and nationality, both spatially bounded. But citizenship can be defined in terms other than those associated with national identity and legally defined rights, duties, and obligations. Citizenship "can also be applied in a wider sense, to the way an individual perceives and practises being a human being in a society, and/or as a member of an interest group, according to his or her own perception of what is right. That is, on the basis of moral judgement" (Williams and Humphrys 2003, p. 4). Beneath this simple distinction, the constitutional and the communal, lies much complexity.

The turmoil in the Middle East, focused on the troubles of Iraq and Syria, exemplifies the difficulties surrounding the definition of national boundaries. The societal pluralism found in metropolitan cities worldwide exemplifies the problems of defining society. For the latter, it can be argued, that the multiplicity of social groups within a specific place yields multiple definitions of citizenship. Thus, citizenship is perceived differently by persons according to, among others, their age, gender, class, religion, race, language, ethnicity, physical and mental ability and location. Just to complicate things further, there is increasing attention being paid to such concepts as world citizenship, transnational citizenship, dual nationality, environmental citizenship and, even, academic citizenship. Some of these can be seen clearly in current crises in Europe, not least in Ukraine, but also in the migration of European citizens within Europe and non-Europeans trying to cross the Mediterranean and the Channel. For ordinary citizens in a stable political culture, none of these issues occupy much of their time and thinking. By contrast, in places of economic hardship, day to day survival is all consuming. Obtaining clean drinking water and an adequate supply of food and firewood is the constant preoccupation of many people, especially in Sub-Saharan Africa. The precariousness of life for persons in places threatened by such natural hazards as severe droughts, riverine flooding, tsunamis, earthquakes and volcanic activity as well as the exigencies of civil strife and war, also overwhelms any concerns with citizenship. The responses to such human disasters by local communities demonstrate citizenship in action. Such active citizenship is often seen to be more effective in the short run than any governmental efforts. All of this is some distance from political discussions about participatory citizenship.

Particular Issues

Terrorism

International terrorism has become one of the most serious, if not *the* most serious of global problems. The ability of terrorist groups to utilise ICTs for propaganda purposes, for planning and implementing activities worldwide and to recruit newcomers to their cause has challenged governments to devise counter-measures that also employ ICTs. Electronic surveillance is part of this, though its use is contentious since the methods employed can be seen as intrusions into private life and there is much scope for misuse. In some countries there is an ongoing debate about the need to censor the internet in an attempt to curb the flow of information that might benefit terrorist groups and this too is contentious.

Gender

Gender equality is one of the Millennium Goals identified by the United Nations for a programme that started in 2000 with completion in 2015. The goal was to promote gender equality and empower women. High on the list of issues prioritised by the UN are the ending of violence and harassment towards women, the equalising of educational opportunities, especially in primary schools, greater participation in political institutions and more involvement of women in political life. The plight of women and how this is handled in the media is exemplified by the attack on Malala Yousafzai in Pakistan in 2012 and the abduction in 2014 by gunmen of more than two hundred and fifty schoolgirls from a Government Secondary School in Nigeria.

In the context of citizenship and governance, at an international institutional level the gender equality agenda is framed by improving the presence of women. At street, village and community level the agenda is framed by finding ways for women's voices to be heard. As a UN Women report asserts that, for gender relations to be transformed, women and girls ... "should see the expansion of the full range of human capabilities and have access to a wide range of resources on the same basis as men and boys, and they should have a real presence and voice in the full range of institutional fora ..." (UN Women 2013, pp. 3–4). For women who own and control mobile phones and computers, their world has been transformed. Provided the technological infrastructure is in place and costs are not prohibitive, they have been empowered to engage in social, cultural and political communication that can lead to various forms of engagement, especially political engagement. They can seek to effect policy changes at local and national levels through formal consultation processes and direct action. Using social media they can engage in international communication, provided that there is a shared language between the correspondents. This is not to suggest that opportunities are equal even within national boundaries. Networking is much easier and quicker in cities than in rural areas, not least because ICTs are more accessible there.

Statelessness and Migration

Statelessness is a contemporary phenomenon that has several facets. For example, there is the statelessness of elite jetsetters, corporate businessmen and women who migrate from home to home and company to company in locations scattered worldwide. They are in the same category as high value sportsmen and sportswomen, and others from the entertainment communities who appear to flit with ease from one continent to another. By contrast, there are the refugees, asylum seekers and migrants whose intracontinental and intercontinental travelling is anything but easy. Regarding the latter, the United Nations High Commission for Refugees estimated that in 2014 some 59.5 million people were forcibly displaced. Of these, 19.5 million were refugees, 38.2 million were internally displaced and 1.8 million were asylum seekers (UNHCR 2015). To bring some scale to these enormous figures, it was estimated that 42,500 people were forced to migrate per day as a result of civil conflict and persecution. ICTs have converted the statistics into personal stories, bringing a sense of immediacy to the forces that have driven people to migrate on hazardous journeys sometimes across oceans and sometimes overland. The mass media project images in real time of scenes from the Mediterranean, the Pacific and the US-Mexican border. They also display similar images from pressure points in Africa and Asia. It is important to recognise that Turkey received most refugees in 2014 (1.59 million), followed by Pakistan, Lebanon, Iran, Ethiopia and Jordan. In Sub-Saharan Africa, Africa's numerous conflicts, including in Central African Republic, South Sudan, Somalia, Nigeria, Democratic Republic of Congo, together produced immense forced displacement totals in 2014, on a scale only marginally lower than in the Middle East. The political consequences of these migrants are immense for the migrants themselves, especially for those who end up in vast camps, as well as for the governments that have to cope with the arrivals when their own resources are inadequate. Integral to the everyday life of the displaced and stateless is the loss of any rights to education, health care, housing and other aspects of life taken for granted in more advantaged communities.

Conclusions

In the context of citizenship and governance, there is an abundance of international documents produced by agencies such as the United Nations, the World Bank and the OECD that have produced discussion papers, organised conferences and published discussion papers, expert reports and published recommendations, roadmaps and sets of goals for national governments to consider. In parallel, national and local governments have addressed these issues for their citizens. Meanwhile, non-governmental organizations, operating at a variety of scales and scholars from a range of academic disciplines have set their own agendas for study, research and communal activity. There are many voices to be heard and ICTs have provided the means for individuals to contribute at many levels. Voices may or may not have power and authority. Their capacity to influence governmental policies and practices will vary according to local and national circumstances. One needs only to consider the resources and opportunities available in contrasting environments such as the megacities of the First World and the scattered rural villages of the Third World. Increasingly, ICTs are creating new inter-linked communities varying from the local to the global. They challenge our conventional understanding of citizenship and pose problems for governments seeking to function efficiently, transparently and accountably.

Open Access This chapter is distributed under the terms of the Creative Commons Attribution 4.0 International License (http://creativecommons.org/licenses/by/4.0/), which permits use, duplication, adaptation, distribution and reproduction in any medium or format, as long as you give appropriate credit to the original author(s) and the source, provide a link to the Creative Commons license and indicate if changes were made.

The images or other third party material in this chapter are included in the work's Creative Commons license, unless indicated otherwise in the credit line; if such material is not included in the work's Creative Commons license and the respective action is not permitted by statutory regulation, users will need to obtain permission from the license holder to duplicate, adapt or reproduce the material.

References

Office of the e-Envoy, Cabinet Office 2002. *In the Service of Democracy: a consultation paper on a policy for electronic democracy.* London: HM Government.

Peters, J. and Abud, M. 2009. E-Consultation: enabling democracy between elections. *IRPP (Institute for Research on Public Policy) Choices*, 15(1): 2-31.

UN Economic Commission for Africa 1996. *African Information Society Initiative: An action framework to build Africa's information and communication infrastructure.* Addis Ababa: UN Economic Commission for Africa.

UN Economic Commission for Africa 2008. *African Information Society Initiative (AISI): A decade's perspective.* Addis Ababa: UN Economic Commission for Africa.

UNHCR 2015. *Global Trends Report: World at War: Forced Displacement in 2014.* Geneva: UNHCR

UN Women 2013. *A Transformative Stand-Alone Goal on Achieving Gender Equality, Women's Rights and Women's Empowerment: Imperatives and Key Components.* New York: UN Women.

Williams, M. and Humphrys G. eds., 2003.*Citizenship Education and Lifelong Learning: Power and Place.* New York: Nova Science Publishers Inc.

Part II
Examples of Global Diversity

Changing Cultures: Changing Lives—Mobilising Social Media During a Health Crisis

Martha Kamara

Abstract Though developing continents such as Africa continue to be challenged by the prevalence of certain health related matters, the emergence of Information and Communication Technologies seems to be promising in managing and monitoring a number of heath related diseases. The rapid growth of mobile phones, computers, and other social media devices in almost all cities and rural areas in Africa has been the catalyst for this change. The use of m-Health and e-Health care strategies developed in tandem with industrialised countries has increasingly contributed to improvements in healthcare. Illustrative are m-Health applications in Africa with particular reference to its use during the recent Ebola crisis in Sierra Leone.

Keywords Healthcare · Digital information sharing · Ebola

Introduction

The rapid development and growth in digital technologies has significantly transformed global and local patterns of communication and dissolved continental and regional boundaries. These innovations have included the World Wide Web, mobile phones, and online social networks such as Facebook, WhatsApp, twitter, LinkedIn, and much more. Social media for example, has transformed and enhanced how families and friends communicate with each other. While a digital divide continues to exist between global north and south, social media has gained currency and popularity at a faster rate than expected in Africa, reaching rural and most remote villages. Most people can now boast of having a mobile phone regardless of where they live and work. The rapid response of social media and other forms of digital communication in Sub-Saharan Africa has seen a steady stream of innovation and

---

M. Kamara (✉)
School of Education, La Trobe University, Melbourne, Australia

© The Author(s) 2016
M.E. Robertson (ed.), *Communicating, Networking: Interacting*,
SpringerBriefs in Global Understanding, DOI 10.1007/978-3-319-45471-9_4

advancement in health, education, governance, and economic development. Communications technologies have redefined and revolutionised traditional communication protocols and cultural practices. Social media provides a fast, cost-effective, and more transparent communication alternative which overcomes limitations that hindered progress in the past. Sectors such as banking and finance have been swift to embrace and exploit digital technologies extensively to improve service delivery. In almost every remote part of Africa, financial transactions are executed across the globe securely and rapidly through one form of technology or the other. Most families with relations around the world can now send and receive money through a simple mobile phone by financial institutions and agents in countries that provide such services.

Notably, the health sector has realised a rapid growth in the use of digital technologies in recent years (Bruce 2002; Khan et al. 2010; Thinyane et al. 2010). Considerable interest in the use of information and communication technologies for health-care delivery has significantly transformed the administration and management of health service delivery globally. Essentially, there has been significant improvements in responding to health service delivery through e-health, a cost-effective and secure use of information and communication technology for health and health-related purposes through the use of various applications and tools (Cole-Lewis and Kershaw 2010).

Industrialised countries such as Australia, USA, Canada, and the UK are now relying on electronic medical records (EMRs) in its various forms to provide quality health care efficiently and effectively. At a universal level, there has been widespread support for an integrated approach to activate global e-health (Blaya et al. 2010) paying particular attention to developing countries in Sub-Saharan Africa. The case of adopting these technologies has been widely embraced though not without enormous challenges as a partial solution to improving universal healthcare with developing countries the most vulnerable due to cultural and financial constraints (Edejer 2000). In an increasingly digital world, prompted by technological advances, economic investment, social and cultural changes, there is growing recognition that inevitably the health sector must integrate information communications technology into its practices if universal healthcare is a priority for world leaders as pronounced by the United Nations and its key agency, the World Health Organisation. This applies whether the goal is to reach all citizens with high quality, equitable and safe care, or to meet obligations for public health records and research.

The African Context

Rapid readiness to embrace the use of communication technology with the potential to improve development projects in Africa, has witnessed significant improvements in facilitating clinical and managerial decision-making in some parts of the African region thanks to the work of health agencies and other non-government organisations such as United Nations Development Program, and World Health Organisation (UNDP 2015a, b). Africa has slowly migrated to e-Health solutions from storage of patient information which traditionally involved stacks of manila files and handwritten notes, to diagnosing, treating, and monitoring of certain diseases such as HIV/AIDS (Ediger 2000) with penetration of digital devices. Being part of a global village has substantially increased and created new opportunities in Sub Saharan Africa though significant challenges are acknowledged. For example, an equitable share of resources in developing countries remains a major challenge in terms of affordability and reception. Migrating to e-Health has so far been of immense benefit though the level of development in technology is remains low, and coverage of digital health technologies in large economies such as Nigeria is still in a neophyte stage (Batta et al. 2015). For the continent, the majority of the population is living in abject poverty with little access to healthcare for medical assistance. Countries such as Kenya, Ghana, Nigeria, Rwanda, Ethiopia, Uganda and South Africa are slowly migrating to e-health facilities in an attempt to facilitate National Health Service delivery in response to achieving United Nations Millennium Development Goals (MDGs) and match the resolution of the World Health Assembly urging nation-states to embrace e-health in an effort to improve and strengthen health systems (Asamoah-Odei et al. 2007).

In Kenya for example, text messages have been used to provide services to HIV/AIDS patients thanks to the rapid spread of mobile phone use. In Nigeria, it is reported that phone companies are providing frequent free text messages on HIV prevention to all subscribers. In South Africa a number of measures have been taken to establish e-health practices in the treatment of diseases such as HIV/AIDS and zoonotic diseases in remote communities. Networks of e-Learning and telemedicine have been established in other countries such as Mali, Senegal, Cameroon, Burkina Faso and Niger in collaboration with some industrialised nations such as France and Switzerland. Such cooperation has been noted with other African countries in promoting transnational e-Health and Telemedicine (Asamoah-Odei et al. 2007).

Sierra Leone—A Case Study

The recent Ebola crisis has had a profound impact on Sierra Leone. At the time of the outbreak there were severe shortages of doctors with only two doctors per 100,000 people (Momodu 2014). Admittedly, the health care system in the country was considered weak, lacking good laboratory facilities for early detection and diagnosis.

Sierra Leone a multi-ethnic country located in West Africa with an area of 71,730 km^2 and a population of approximately 6 million people gained independence from British indirect rule in 1961. According to the Human Development Index in 2011, it is one of the world's poorest nations ranking 180 out of 187 countries. English is the official language. Krio is widely spoken by 90 percent of the population and represents the dominant lingua franca of ethnic groups. Poverty is widespread with more than 60 percent of the population living on less than $1.25 a day (UNDP 2015a). Illiteracy and unemployment are very high. Despite social, economic and political problems, there is steady capacity rebuilding in collaboration with aid agencies such as UNDP in mobilising the nation's opportunities since the end of the civil war in 2002 (UNDP 2015b). Along with infrastructural rebuilding across the nation, mobile communications technologies have become central to programs for change. Although mobile phone companies are established in major cities, the population has embraced technology at a faster rate than expected including a significantly illiterate population living in rural areas and villages. Local village traders are now empowered to transact business directly with customers by minimising third party reliance on external agencies.

Sierra Leone and Modern Telecommunication

Paradoxically, while communication gap between governments and communities undermined the efficacy of the emergency response, closely linked to weak national capacities overall, especially in terms of ensuring access for all to basic services for health, water, sanitation, education and social protection, more people in Africa have access to mobile phones. In Sierra Leone, mobile phone usage has increased tenfold with only about 113,000 subscribers; however by 2008, there were over 1 million subscribers (Petifor 2011). The most prominent users are young people between 15 and 35 years. It is reported there are about 2 mobile phone users per household in the rural areas (thanks to the provision of Solar photovoltaic (PV) which provides a solar mobile phone charging system (Mansaray 2013). The promptness and capacity of instantaneous communication via social media, the internet and Facebook in this instance of crisis brought to bear the ways in which traditional social and cultural practices relating to communication protocols were being transformed. Populations within Sierra Leone are no longer isolated from the world. Social media provides a means to bypass traditional knowledge cultures and practices and opens up possibilities for widespread mobilisation of news and information through communications technologies. Nonetheless, it has taken a crisis in the health sector such as Sierra Leone and other developing countries in Africa to move e-Health from the periphery to the centre of strategic health planning.

The Ebola Crisis

While there are distinct social and economic differences between affluent city residents, villages, and transient slum dwellers, the Ebola crisis demonstrated an instance where social media became a significant communication medium for reaching populations within Sierra Leone, for communicating to the global world and the Sierra Leonean diaspora.

The Ebola Virus Disease (EVD) crisis highlighted the fragility of Sierra Leone's health, education and social structures when faced with the rapidity of disease. As noted by the President of Sierra Leone at the recent Ebola Conference held in Brussels, noted that at the outbreak of the virus, 3100 citizens died which included over 400 children and hundreds of orphans. This situation decimated an already crumbled economy with heavy casualties in the education and health sectors. Schools, colleges, universities, hospitals, clinics, commercial centres went into lock down. There was widespread flight of residents from cities, towns, and villages in fear of being contaminated even though most residents did not fully understand the ramifications of the outbreak. Sadly, a crippling government most willing to protect and secure its residents did not have immediate answers nor the resources to manage the enormity of the situation. Frontline health workers became casualties including a few specialised physicians. The EVD made no distinctions between social status, wealthy or poor, village or city. The population was extremely vulnerable, fearful and anxious about where and when, and who the disease would take. Unfortunately, the crisis took the entire world by surprise and a rapid response to treatment and management was slow, fragmented, and challenging.

The health crisis exacerbated the fragility of a nation recovering from a 10 year brutal civil war in which 50,000 people were killed, infrastructure destroyed, and more than 2 million residents displaced. Thousands of professionals left for Europe, America, and Australia, leaving a significant intellectual, social, economic vacuum to overcome for rebuilding the nation's political system, civic infrastructure, health, education, and the economy. Health and education are two critical areas where much help is needed to meet the UN Millennium Development Goals of eradicating poverty, hunger and disease and for developing more resilient nations. The challenges of power shortages was one of the major catalysts in casualties that decimated the country. For example, speaking on a mobile phone a dying woman's last words to a relative from her Ebola bed were "I need to desperately charge this mobile phone but there is no power and I have been abandoned here by the health workers". Ultimately, her voice faded as the phone shut down. Shortage of protective equipment, ignorance of the outbreak, and cultural practices hampered prompt treatment of patients.

Despite the spontaneous outbreak, international agencies in a coordinated effort eventually mobilised to reach out and intervene with high tech equipment and management strategies, in cooperation with the entire nation. For example, UNDP offices were upgraded in the three hardest hit countries in West Africa—Liberia, Guinea, and Sierra Leone to provide videoconferencing kits to boost internet bandwidth worth over a million dollars. The provision of this facility made all the difference in coordinating efforts of staff in the affected countries and industrialised nations where medical staff were located. In part, the widespread use of social media and other forms of telecommunication became one of the catalysts for combating the disease though there was less readiness for telehealth services. It became expedient for a paradigm shift in an integrated approach to re-establish institutional frameworks with greater adoption of digital technologies. One of the critical steps taken at the time was the proper establishment of telecommunication channels and training of health service staff in the treatment and management of a crisis of that magnitude. To facilitate and harmonise communication nationally, Ebola centres were established in certain regions of the country and the internet became a reliable facility at Ebola Response Centres (ERC) managed by e-Health at the "Alert Pillar" unit of the Ebola Response Centre or C Command Centre in major cities. The Alert Pillar was assigned the responsibility for receiving all alert calls (sick, death, suspect, and security). These calls were then redirected to the appropriate response wings of various unit—dignified burial, surveillance, life case, psychosocial and social mobilization, and Security units for necessary action. The availability of the internet greatly facilitated administration and management between the government and agencies in speeding up communication between districts when recording information on the number of outbreaks and death notices. The setting up of a database was helpful on many fronts in assisting access to information that communities would not normally receive if relying on traditional technologies of communication (for example, fixed telephone, mail, newspapers). Having up to date accounts around the country assisted government officials and medical staff to understand patterns of disease by having reliable statistics on actual and approximate figures.

The crisis illustrated the critical need for communication efficiency to reach populations in educating them about health and disease prevention. Mobile technologies and internet access indicated fundamental shifts in managing crises of this nature. There was a ground breaking innovation of mobile payments to Ebola staff, cloud computing and open source information management systems to efficiently manage the crisis with the UNDP as frontline agency (WHO Ebola Response Team 2014). During this time there was an exponential increase in the use of social media, from villages to cities, and to the broader world. Sierra Leoneans themselves utilised social media to convey news, find relatives and friends and post updates on the progress and status of individuals as the disease unfolded in time. Already grappling with a struggling economy for Sierra Leone the socio-economic impact will be lasting. It will need a global effort to re-establish education, health, and other social services in the country. The unflinching efforts of the Sierra Leone government to provide universal health care has been dealt a blow.

The Way Forward

While it may take a number of years to rebuild the struggling economy, good strategies will progress the efforts of international bodies and the government. Essentially, some long term investment and commitment is required from all key stakeholders. Global resilience, preparedness, and response have been widely criticised (Brown and Cropley 2014). Lessons learned in the hardest possible way are essential starting points for reforming recommendations from the World Health

Organisation (WHO) in implementing policy initiatives from the country. Fundamentally, education is a key target in the recovery process. Basic primary health care training is necessary. Health care staff including community health care workers need to work collaboratively to maximise the impact of containing the disease; yet even is challenging since affected countries are grossly understaffed. It is imperative that the presence of the United Nations and other international agencies remain is a priority even when a zero target had been reached. As a developing country, it is important that cultural practices that have been long established and which may have a potential to counter recovery practices are dealt with in a sensitive manner. A dialogic process with community people will yield better answers and empowerment than coercive measures. The outbreak has created high rates of unemployment particularly among the youth. The implementation of vocational education programs could be the answer to engaging this cohort. Consequently, it will need a genuine effort from the government and citizens to ensure that resources provided for the recovery process are utilised accordingly.

Conclusion

Though wireless technology has altered health care delivery, there remain enormous possibilities in responding to health epidemics particularly in emerging countries. With the widespread use of mobile technology and social media platforms in Africa, there is opportunity to capitalise on e-Health and e-Education modes of delivery that should incorporate culturally appropriate ways of accessing and interacting with such technologies. Harmonising education and health systems through cultural and social interfaces along with the development of specific applications for mobile phone use may provide a way forward to overcome significant challenges presented at the time of the crisis. No doubt there are important lessons learned at various levels nationally and internationally. With the widespread use of mobile phones and other telecommunication devices, a new direction in enhancing the provision of cost-effective devices and services will play a major role in e-Health service delivery not only for diseases such as Ebola but other life threatening diseases.

Open Access This chapter is distributed under the terms of the Creative Commons Attribution 4.0 International License (http://creativecommons.org/licenses/by/4.0/), which permits use, duplication, adaptation, distribution and reproduction in any medium or format, as long as you give appropriate credit to the original author(s) and the source, provide a link to the Creative Commons license and indicate if changes were made.

The images or other third party material in this chapter are included in the work's Creative Commons license, unless indicated otherwise in the credit line; if such material is not included in the work's Creative Commons license and the respective action is not permitted by statutory regulation, users will need to obtain permission from the license holder to duplicate, adapt or reproduce the material.

References

Asamoah-Odei, E., de Backer, H., Dologuele, N., Embola, I., Groth, S., Horsch, A., Ilunga, T., Mancini, P., Molefi, M., Muchenje, W., Parentela, G., Sonoiya, S. Squires, N., Youssouf, M., and Yunkap, K. 2007. E-Health for Africa: Opportunities for enhancing the contribution of ICT to improve health services. *European Journal of Health Services*, 12(1), 1-38.

Batta, H., Udousoro, N., & Abubakar, Y. 2015. Digital health technologies and implications for developing country media and health communication. *New Media and Mass Communication*. 41, ISSN 2224-3267 (Paper) ISSN 2224-3275 (Online). www.iiste.org

Blaya, J., Fraser, H., and Holt, B. 2010. E-Heath technologies show promise in developing countries. *Health Affairs*, 2, 244-251.

Bruce, J. 2002. Marrying modern health practices and technology with traditional practices: issues for the African continent. *International Council of Nurses, International Nursing Review*, 49, 161-167.

Brown, C., and Cropley, I. 2014. Ebola virus disease: where are we now and where do we go? *Postgrad Medical Journal* November, 90(1069). 610-12. Editorial.

Cloe-Lewis, H and Kershaw, T. 2010. Text messaging as a tool in disease prevention and management. *Epidemiologic Reviews*, 32, 56-69.

Edejer T. 2000. Disseminating health information in developing countries: The role of the internet. *BMJ*, 321, 797–800.

Khan, J., Yang, J., and Hahn, J.S. 2010. 'Mobile' health needs and opportunities in developing countries. *Health Affairs*, 29(2), 254-261.

Mansaray, K. 2013. United Nations Industrial Development Organisation (UNIDO) Solar Lantern project in rural Sierra Leone: Clean tech for green industry. *UNIDO Solution Forum at GSSD Expo*.

Momodu S. 2014. Ebola: fighting a deadly virus. *Africa Renewal*, December, p.12-14.

Petifor, A, 2011). Understanding communications in Sierra Leone. *Mamaye, Evidence for Action*, September

Thinyane, H., Hansen, S., Foster, G., and Wilson, L. 2010. Using mobile phones for rapid reporting of Zoonotic diseases in rural South Africa. In Global Telehealth: Selected Papers from Global Telehealth. A. Smith and A. Meader (Eds). IOS Press, Amsterdam.

UNDP, 2015a. Restoring livelihoods and fostering social and economic recovery: UNDP response to the Ebola crisis in Sierra Leone. UNDP www.undp.org/.../Ebola%20Docs./

UNDP 2015b. Hi-tech conferencing helps UNDP coordinate anti-Ebola action. www.undp.org/.../Ebola%20Docs./

WHO Ebola Response Team 2014. Ebola virus disease in West Africa: The first nine months of the epidemic and forward projects. *The New England Journal of Medicine*. 371(16), 481-95.

Bridging the Digital Divide: Everyday Use of Mobile Phones Among Market Sellers in Papua New Guinea

George N. Curry, Elizabeth Dumu and Gina Koczberski

Abstract Access to mobile technologies is transforming the daily lives of poor subsistence farmers in Papua New Guinea. However, the success of this access depends on infrastructure and where connectivity is poor there is evidence of a digital divide. Nevertheless, increasing affordability of internet access is helping to bridge the development gap.

Keywords ICT revolution · Developing world · Gender divide · Transformation

Introduction

Papua New Guinea, like many Pacific Island nations and most of the developing world, is experiencing an ICT revolution as access to information and communication technology (ICT) infrastructure expands rapidly (Cave 2012). This ICT revolution has the potential to revolutionise areas like agricultural extension (E-Agriculture), Health (E-Health), Banking (E-Finance) and Education (E-Education) (Cave 2012; Maumbe 2013; PRIF 2015). Ownership of mobile phones is expanding rapidly and smart phones are putting internet technology into the hands of poor subsistence farmers following a way of life largely outside the market economy (Curry and Koczberski 2013). The rapid uptake of ICTs by the poor can be attributed to greater affordability, accessibility, and adaptability of ICTs (McNamara et al. 2011; GSMA 2015).

The recent wave and adoption of new technologies and digital applications across the globe has been likened to a "Great Transformation" (Fuchs 2013, p. 16) that is fostering rapid growth in accessing information and new ways of producing,

---

G.N. Curry (✉) · E. Dumu · G. Koczberski
Curtin University, Bentley, Australia
e-mail: G.Curry@exchange.curtin.edu.au

creating, sharing and communicating knowledge. It is suggested this revolution is at least as transformative as the Green Revolution technologies once heralded as the panacea for world hunger as they were rolled out across the developing world in the 1960 and 1970s (Fuchs 2013). This remarkable transformation is at once liberating and socially and economically empowering, with the potential to transform gender relations and bridge the economic and social divides within and between countries.

However, whilst there has been an enormous increase in ownership of, and access to, ICTs worldwide, the growth across the globe has been uneven. The developing and least developed countries lag well behind developed industrial countries in engaging in the digital age. Although the majority of people accessing the internet live in poor countries, per capita use is much lower than in developed nations. About 21 percent of the population of the developing world have internet access compared with around 84 percent of the developed world's population (ITU 2014). For example, while access to the internet has increased in PNG since the introduction in 2011 of a mobile broadband service and the expansion of high-speed 'third generation' (3G) and 4G mobile broadband networks, only 9 percent of the population have access to the internet (ITU 2014; see also Cave 2012; Logan 2012).

The digital divide is not only between countries, but within countries. Despite more rural people gaining access to ICTs in poorer countries, there remain large differences in the extent of network coverage between rural and urban areas. In a study of 17 Sub-Saharan countries in 2010, 69 percent of urban respondents owned a mobile phone compared with only 53 percent of rural respondents (Totora and Rhealt 2011). Similarly, a 'gender digital divide' is typical of many developing countries where women and girls have less access to ICT than men and boys (GSMA 2015). These divisions in the ownership of and access to ICTs further marginalise the rural poor and women from the benefits of social and economic change resulting from the expansion of ICT availability.

This chapter examines the extent to which a digital divide exists in PNG. The chapter reports on preliminary research on the recent uptake of mobile phones to investigate whether service and economic divides between urban and rural PNG and an entrenched gender bias in PNG are also reflected in new digital divides. The question this chapter addresses is whether mobile phone technology is bridging existing gender and spatial inequalities or simply reinforcing them. The chapter begins with a brief overview of spatial and gender inequalities in PNG and then considers how mobile phone technologies are being taken up and used by men and women in both urban and rural/remote locations. The data for the chapter are drawn from several research projects in which the authors are engaged: (1) fruit and vegetable sellers at informal markets in Mt Hagan, the capital of Western Highlands Province (WHP) and Kokopo, the capital of the island province of East New Britain (ENBP); (2) cocoa growing households in ENBP, Milne Bay (MBP) and the Autonomous Region Bougainville (ARB); and (3) oil palm smallholders in West New Britain (WNBP) and Oro provinces (OP) (Fig. 1).

Fig. 1 Papua New Guinea

Papua New Guinea

PNG has experienced a mining and mineral boom over the past few decades (UNDP 2014), with GDP growth rates averaging above 6 percent for over ten years (ADB 2014). Despite this mineral-driven growth, poverty remains widespread with the majority of the population, especially the rural population, benefiting little from this growth. Approximately 87 percent of the national population of over 7 million live in rural villages and are highly dependent on agricultural-based activities to meet their everyday social and economic needs. Most people rely on their own food gardens for everyday consumption and for income, and there are very few cash-earning opportunities outside of agriculture. Nearly 90 percent of cash income in rural areas is from export cash crops such as cocoa and coffee and the local marketing of food crops and betel nut (Allen et al. 2001; Sharp 2012), with over two-thirds of this income from export cash crops.

The Rural-Urban Divide

The physical geography of PNG presents major challenges to the provision of basic infrastructure and services. Approximately 52 percent of the total land mass in PNG is classified as mountains and hills (Allen et al. 2005). Access to roads, health, education, telecommunication networks, and markets by rural villagers, especially

Plate 1 Coffee often has to be carried long distances to market in the PNG highlands

those isolated by rugged terrain (Plate 1) is very difficult and costly to the point that livelihood options are severely curtailed. Often migration to urban and rural resource development sites is the only viable option for people from these communities. Indeed, migrants from remote, poorly serviced and disadvantaged rural areas and small islands constitute a high proportion of the growing urban population (Koczberski et al. 2001; Storey 2010; Curry et al. 2012; Numbasa and Koczberski 2012).

The rural-urban divide is very marked in PNG and is apparent across a wide range of variables including cash income levels, food security, education and literacy and a whole suite of health indicators. The UNDP (2014) point out that while an urban/rural divide in human development is common in many parts of the world, it is particularly stark in PNG. For example, despite fairly good access to customary land for most of the rural population, food poverty at 28.5 percent of the rural population has been estimated to be double the urban rate (NSO 2011). The poorer quality diets in rural areas are also reflected in adverse nutritional outcomes. Nationwide household surveys in 1996 found almost half of rural children were stunted compared with one-fifth of urban children (Gibson 2000). Rural people also have less access to primary health care compared with the urban population, and rural children are more likely to carry a greater burden of infection because of their poor access to good health care (Gibson 2000; Howes et al. 2014). Similarly, illiteracy is more prevalent in rural areas. At 40 percent, the rural illiteracy rate is three times the urban rate (13 percent) (NSO 2011; Kare and Sermel 2013). The considerable gap in literacy rates between urban and rural PNG reflects not only the poorer access to schools in rural and remote areas, but also the difficulty for schools in maintaining staffing levels to teach literacy and numeracy skills (Waffi et al. 2015). Such low rural literacy rates and the very different education levels between rural and urban PNG also help explain the lower nutritional status of rural children (Gibson 2000).

These service and income inequalities also apply to material conditions of living. In housing, for example, 72 percent of rural houses have walls constructed from bush materials compared with 7 percent of urban houses (NSO 2011). Roofing iron is highly prized in PNG and is considered to be an indicator of wealth and status. In rural PNG, 29 percent of houses are roofed with corrugated iron compared with 89 percent of urban homes (NSO 2011). In terms of durable household goods, the same pattern is repeated with urban households having much higher ownership rates of consumer goods. For example, almost half of urban households have a stove compared with just 4 percent of rural households; 46 percent have a refrigerator while less than 3 percent of rural households do so (NSO 2011).

The Gender Divide

PNG's constitution, written in 1973, has a provision for equal opportunities for all citizens. Despite its noble intentions, and PNG being signatory to several international conventions on gender equality (e.g., Convention on the Elimination of all Forms of Discrimination against Women—CEDAW), discrimination against women and girls is pervasive. The Gender Inequality Index (GII) ranks PNG 134 out of 148 countries (GoPNG 2013), and gender inequalities in PNG have proven extremely hard to address. On virtually every socio-economic and health indicator, women fare worse than men (GoPNG 2013), and the country is one of the few nations in the Asia-Pacific region yet to achieve gender equality at primary school level. Rates of gender-based violence are amongst the highest in the world (World Bank 2012a; GoPNG 2013; UNDP 2014): two-thirds of women are estimated to have experienced gender-based violence (GoPNG 2013). Women in PNG not only risk high rates of violence, but also have fewer economic opportunities than men, have high maternal mortality rates (over 100 times higher than the rate for Australia—UNICEF 2013) and shorter life expectancy than men. They also experience inequalities within the home which limits their access to household income and participation in decision-making (Overfield 1998; Wardlow 2006; Koczberksi 2007; Macintryre 2008; World Bank 2012b; UNDP 2014).

Education statistics also show a strong gender disparity, despite recent improvements (NSO 2011; UNDP 2014). PNG women are more likely to be illiterate, have lower levels of primary and secondary school attainment, and to be less represented (38 percent) at university level than men (e.g. Gannicott and Avalos 1994; Gibson and Rozelle 2004; ADB 2012; DFAT 2012; Kare and Sermel 2013; UNDP 2013). A higher proportion of males than females can read and write (69 percent compared to 57.3 percent) and just 6.8 percent of adult women have secondary or tertiary level education compared with 14.1 percent of men (UNDP 2013).

The Introduction of Mobile Phones to PNG

The mobile phone was introduced in 2003 into a country where telecommunication networks are relatively recent. The first radio and telephone technologies were installed in Rabaul, ENBP in 1907 (Fig. 1) and over the following two decades they were extended to other major centres of the country (Sinclair 1984; Suwamaru 2013). In 1933, PNG established its first government information radio station and by 1975, when PNG gained political independence, the telephone network was accessible only to a small group of the urban elite (Ogden 2013). Fixed line telephones remained out of reach of most of the population and in 2007 PNG had one telephone per 100 inhabitants; the lowest penetration rate among the Pacific Island nations (Ogden 2013; Crocombe 2001).

With limited telecommunication networks up until the introduction of mobile telephony, most information was relayed locally using traditional communication devices such as conch shells, drums, yodelling and slit-gongs (*garamut*). Radio and postal services were commonly used for communications over longer distances (Suwamaru 2013). As Telban and Vavrova (2014, p. 3) point out, prior to mobile telecommunications in PNG, "[r]emote villages were relying mainly on letters and *tok save* 'announcement, notification' over the radio or on traditional means of carrying the messages by either word-of-mouth or a drum signal".[1]

When mobile phone services were introduced, coverage was provided by the state-owned Telikom PNG monopoly and limited to major urban centres like Port Moresby, Lae, Madang, Goroka and Mt Hagen (DFAT 2004). In 2006 there was one mobile phone subscription per 100 people (World Bank 2012c). In 2007 mobile phone access improved considerably when the government removed the Telikom PNG monopoly on fixed line and mobile phone services, allowing new providers to enter the market. There are now three mobile phone providers in PNG. One of the first new players, Digicel, entered the market in July 2007. Within eight months of Digicel's arrival, coverage had expanded from two to ten per cent of the population (Watson 2011), and by 2010 it had reached 28 percent of the population (World Bank 2012c). This initial rapid uptake of mobile phones is reflected in the data collected for WHP and ENBP, with most of the uptake occurring within two years of the arrival of Digicel services (Fig. 2). Over time, mobile phones have penetrated remote rural villages and outer islands that hitherto had been extremely poorly serviced, and often without electricity, health services, reticulated water or road access. By 2014, 41 percent of the population had mobile phone access (Suwamaru 2015).

---

[1] Radio '*tok save*' is an announcement or message relayed over the radio waves via radio stations. People deliver hand-written messages to the radio station to be read on air to relatives and friends living elsewhere.

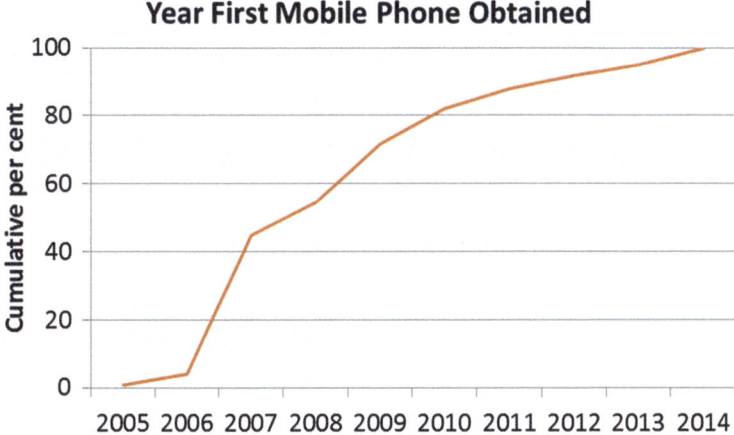

Fig. 2 The uptake of mobile phones in WHP (n = 51) and ENBP (n = 24)

More recently there has been significant growth in the use of mobile broadband to access the internet. This is driving the massive rise in the use of social media in PNG. By November 2015, there were 625,874 internet users (Internet World Stats 2015). Facebook is the most common use of the internet and it is the largest online network in the Pacific with around 700,000 Facebook users in Pacific Island Countries (PRIF 2015). In November 2015 there were over 350,000 registered users of Facebook in PNG, many of whom were aged 18–24 years (Internet World Stats 2015; see also Logan 2012).

The Rural-Urban Digital Divide

Although the uptake of ICT continues to rise in PNG and network coverage continues to improve, preliminary evidence indicates that the so-called 'digital age' remains an elusive dream for many Papua New Guineans, especially those living in remote areas (Cave 2012). In 2010, 39 percent of households in four field sites in EHP owned a mobile phone. Although the data are not directly comparable, more recent data collected from urban markets in WHP and ENBP (sellers from relatively accessible locations), show that almost 100 percent of sellers or someone in their household owned a mobile phone. The same data set revealed that 75 percent of individuals over 18 years of age owned a mobile phone. Mobile phone uptake appears to be reaching saturation point, at least for some urban locations and rural villages with relatively good access to town.

There is, however, a clear spatial divide. Mobile phone ownership rates are considerably lower in remote rural locations than in urban areas or rural villages near town (Table 1). Initially, this urban-rural divide would have reflected mobile phone access and connectivity prior to 2007 when mobile telephony was limited to

Table 1 Mobile phone ownership among export cash cropping households in PNG by remote and accessible locations* (percent of households)

Location	Year**	Accessible locations (percent of households)	Remote locations (percent of households)
EHP	2010	66.5 (n = 195)	12.5 (n = 137)
WNBP	2012	87 (n = 206)	
ENBP	2014	99 (n = 59)	
WHP	2014	99 (n = 63)	
ARB	2015	84 (n = 98)	
ENBP	2015	89 (n = 88)	
MBP	2014		40 (n = 63)

*Accessible means within a half day's travel by road of a major urban centre
**Denotes year of data collection

major urban centres and areas close to town. Data collected in 2010 in the EHP (following the expansion of mobile coverage to rural areas), found that the rate of ownership of mobile phones in villages close to town was five times higher than in remote locations in the province.[2] It is likely that this divide has become less marked through time as mobile phone coverage has expanded to include many remote areas. However, in 2014, on Misima Island, MBP the rate of phone ownership was about half that of the accessible sites (Table 1).

This rural-urban digital divide is more than simply a reflection of the spatial distribution of mobile phone services. As noted above, rural and remote PNG is disadvantaged on many indicators, including health, education and income. In Table 1 mobile phone ownership was high among rural households in WNBP, ENBP and ARB who resided not far from town and had access to a regular income from export cash crops and the sale of fresh food produce at town markets. People living in remote rural areas, such as in parts of EHP and MBP have fewer opportunities to earn an income, and incomes tend to be much lower than in areas accessible to town. For example, in the relatively accessible sites in the EHP study, the average number of income sources per household (excluding coffee) was higher than in the remote sites and the potential to earn high incomes through the commercial production of vegetables was also significantly greater. Thus, acquiring and maintaining a mobile phone (e.g. charging batteries and phone charges) is a much greater financial challenge in remote locations.

Table 1 indicates that PNG, like other poor countries where there has been a large increase in the ownership of mobile phones, is showing signs of an entrenched digital spatial divide. It is possible that mobile phones will exacerbate the already stark rural-urban divide. Undoubtedly, people in rural and remote areas will benefit from new services that mobile phone telephony enables such as banking and possibly an extension of agricultural services. But the potential benefits of the

---

[2]The EHP fieldwork was in four sites: Bena and Asaro were accessible sites not far from Goroka town, and Marawaka and Baira were in remote parts of the province without road access.

technology will be much greater in urban PNG and in villages close to town where people have greater capacity to capitalise on the development opportunities of this new technology. They already earn higher incomes which gives them greater capacity to access information on the internet. Improved market information delivered through mobile phones will enable people living near town to capitalise on these new opportunities while such opportunities will be less likely to be taken up by people in rural and remote areas.

The Gender Digital Divide

A 'gender digital divide' where women have less access to mobile phones than their male counterparts is also found in PNG (GSMA 2015). Watson (2011) found that men in Megiar, north of Madang in PNG, were more likely than women to own a mobile phone. This gender difference in mobile phone ownership was not detected at Watson's second field site of Orora on Karkar Island. If a gendered digital divide were present in PNG, one would anticipate that men would adopt the new technology earlier than women, and they would make more use of the technology because they have more political and economic power (higher incomes) than women.

In terms of early adoption of the technology, the 2014 data collected among market sellers in ENBP and WHP were not clear cut. At both sites the average year of adoption was 2008 for both men and women. However, in ENBP, women tended to adopt mobile phones slightly earlier than men and in WHP the reverse was the case with men adopting mobile phones earlier than women (Table 2). This gendered difference in the uptake of mobile phones between the two provinces reflects, in part, the relative status of women in the two provinces: the Gazelle Peninsula of ENBP is matrilineal and women certainly have more status than women in the strongly patrilineal highland societies of WHP.

Further evidence of this gendered difference in mobile phone telephony between matrilineal ENBP and patrilineal WHP is reflected in rates of mobile phone ownership (Table 3). While men in both provinces have higher rates of phone

Table 2 The average year of when first mobile phone was acquired by gender for ENBP and WHP

	Female	Male	Male and Female
ENBP	2008.40	2008.83	2008.72
WHP	2008.72	2008.48	2008.64

Table 3 Percentages of males and females over 18 years of age owning a mobile phone for ENBP and WHP

	Female	Male	Male and Female
ENBP	70.59	81.39	76.21
WHP	62.18	87.85	74.34

Table 4 Average numbers of outgoing and incoming calls and SMS messages per day by gender for ENBP and WHP

	Mobile phone use (average per day)	Female ENBP	Male ENBP	Female WHP	Male WHP
Calls	Outgoing	1.41	0.79	1.39	2.02
	Incoming	1.03	0.78	1.68	2.27
SMS	Outgoing	0.43	0.1	0.21	0.86
	Incoming	0.4	0.5	0.8	1.17

ownership than women, the gender disparity is more marked in WHP where a strong ideology of male dominance leaves women with less social, political and economic autonomy and power than men. There is a 25.7 percent point difference in phone ownership between men and women in WHP compared with a 10.8 percent point difference in ENBP. Clearly, the relative social and economic status of women in the two provinces is reflected in phone ownership rates.

These gender differences by provinces are also apparent in how the mobile phone is used and how frequently men and women make and receive calls and SMS messages. In matrilineal ENBP, women make and receive more calls than men, whereas in patrilineal WHP the reverse is the case and again the disparity between the genders is starker than in ENBP (Table 4). Similarly, in ENBP, women send more SMS messages than men and receive about the same number as men, while men from WHP send and receive more texts than women (Table 4).

In summary, it seems that mobile phone technology is not yet contributing to a significant erosion of the gender divide in WHP society. However, it is likely that mobile phone technology is empowering women by widening their social networks beyond their immediate communities by allowing them to maintain contact with relatives and friends living elsewhere. This is especially important for women in patrilineal societies when they move to the villages of their husbands and have limited contact with their natal relatives living elsewhere. It is also likely that the technology empowers women market sellers economically through, for example, enabling them to contact relatives in town to determine prices at local markets before committing to transporting garden produce to market. These benefits, of course, pertain more to women in accessible locations not far from urban markets who can respond quickly to such opportunities.

Overall, however, when ENBP and WHP are compared it appears that the uptake and use of mobile phone technology reflects existing patterns of gender inequality. Women in matrilineal ENBP have much higher status in their society and this is reflected in their higher rate of mobile phone adoption and use relative to men, whereas in patrilineal WHP with an entrenched high level of gender inequity, men are clearly benefiting far more from the new technology than women.

Conclusion

While the adoption and spread of mobile phone technology in PNG has been nothing short of spectacular in a 'big bang' expansion and adoption of the technology, the jury remains out in the capacity of the technology to bridge the rural-urban and gender divides which are so pronounced in PNG. Clearly, there are major development benefits from adopting the technology. While men and women residing in remote locations and urban centres will benefit, it is debatable whether the technology will enable a 'catch-up' where women's social and economic status relative to men will increase and people living in remote areas will improve their economic situation relative to urban dwellers. As the data have revealed, the uptake and use of the technology has reflected the relative status of men and women with women in ENBP having greater gender equity in the adoption and use of the technology than their female counterparts in WHP where women's status is much lower than men. One potential consequence of the digital divide is a growing 'knowledge divide' along gender lines and between urban and accessible PNG on the one hand, and rural and remote PNG on the other hand.

Also, because the social and development benefits that mobile phone technology delivers will be greater in accessible locations than in remote locations, mobile phones may exacerbate existing spatial inequalities and further marginalise the rural poor. People in remote areas simply cannot act on the information that mobile phones deliver in the same way that people near town can do. High prices in town markets, for example, mean that a vegetable grower or a coffee producer living near town can act more quickly on that information than a producer in a less accessible location.

Finally, the rural-urban digital divide may also exacerbate the gender divide in mobile phone technology, such that the gender inequalities in ICT access in rural areas are more pronounced than in urban areas. This gender divide overlain and exacerbated by the rural-urban divide is not simply because of greater network coverage and incomes in urban areas, but is also likely to reflect the differential status of women in rural and urban PNG. Women's status in urban PNG has improved with modernisation and development, whereas in rural and remote PNG, traditional gender roles and the relative status of men and women have remained largely unchanged. One would anticipate that the capacity of mobile phones to erode gender divides and improve the status of women will be stronger in rapidly modernising urban centres and villages close to urban centres thereby adding a spatial dimension to the gender divide.

Acknowledgments Data collection on mobile phone ownership among oil palm smallholders in WNBP and cocoa smallholders in ENBP, ARB and Milne Bay was done in collaboration with researchers from the PNG Oil Palm Research Association and the Cocoa and Coconut Institute, PNG, respectively. Data collection in EHP in 2010 was in association with PNG Coffee Industry Corporation. We are grateful for the cooperation of villagers who gave their time to be interviewed.

Open Access This chapter is distributed under the terms of the Creative Commons Attribution 4.0 International License (http://creativecommons.org/licenses/by/4.0/), which permits use, duplication, adaptation, distribution and reproduction in any medium or format, as long as you give appropriate credit to the original author(s) and the source, provide a link to the Creative Commons license and indicate if changes were made.

The images or other third party material in this chapter are included in the work's Creative Commons license, unless indicated otherwise in the credit line; if such material is not included in the work's Creative Commons license and the respective action is not permitted by statutory regulation, users will need to obtain permission from the license holder to duplicate, adapt or reproduce the material.

References

ADB (Asian Development Bank). 2012. Papua New Guinea: Critical development constraints. Mandaluyong City. http://www.nicta.gov.pg/publicinquirynew/%20%20RSD%202nd%20Discussion%20Paper/Digicel%20Response%20ADB%20PNG%20-%20critical%20development%20constraints.pdf. Accessed 15 May 2013.

ADB (Asian Development Bank). Asia Development Outlook 2014. Manila: ADB.

Allen, B., Bourke, R.M. and Hanson, L. 2001. Dimensions of PNG village agriculture. In Bourke, R.M., Allen, M. & Salisbury J. (eds), *Food Security for Papua New Guinea*, pp. 529-553. Proceedings of the Papua New Guinea Food and Nutrition 2000 Conference, ACIAR Proceedings No. 9. Canberra: Australian Centre for International Agricultural Research.

Allen, B., Bourke, R. M. and Gibson, J. 2005. Poor rural places in Papua New Guinea. *Asia Pacific Viewpoint*, 46(2), 201-217.

Cave, D. 2012. *Digital Islands: How the Pacific's ICT Revolution is transforming the Region.* Sydney: Lowy Institute for International Policy, New South Wales.

Crocombe, R. 2001. *The South Pacific.* Suva, Fiji: University of the South Pacific.

Curry, G.N., Koczberski, G. and Connell, J. 2012. Introduction: enacting modernity in the Pacific? *Australian Geographer* 43(2), 115-125.

Curry, G.N. & Koczberski, G. 2013. Development Implications of the Engagement with Capitalism: Improving the Social Returns of Development. In: McCormack, F. & Barclay, K. (eds) *Research in Economic Anthropology, Engaging with Capitalism: Cases from Oceania*, pp 335-352. Bingley: Emerald Group Publishing Limited.

DFAT 2004. Annual Report 2004–2005. http://dfat.gov.au/about-us/publications/corporate/annual-reports/annual-report-2004-2005/downloads/CompleteAnnualReport.pdf. Accessed 12 Jan 2016.

DFAT 2012. Annual Report 2012–2013. http://dfat.gov.au/about-us/publications/corporate/annual-reports/annual-report-2012-2013/pdf/dfat_annual_report_1213.pdf. Accessed 12 Jan 2016.

Fuchs, R. 2013. Introduction Part I, From Heresy to Orthodoxy: ICT4D at IDRC. In: Elder, L., Emdon, H., Fuchs, R. & Petrazzini, B. (eds) *Connecting to ICTs to Development. The IDRC Experience*, pp 1-16. **London: Anthem Press.**

Gannicott, K.G. and Avalos, B. 1994. *Pacific 2010: Women's Education and Economic Development in Melanesia.* Canberra: National Library of Australia.

Gibson, J. 2000. Who's Not in School? Economic barriers to universal primary education in Papua New Guinea. *Pacific Economic Bulletin* 15, 46-58.

Gibson, J. and Rozelle, S. 2004. Is it Better to be a Boy? A disaggregated outlay equivalent analysis of gender bias in Papua New Guinea. *The Journal of Development Studies* 40, 115-136.

GoPNG 2013. *The Future We Want. Voices from the people of PNG. Post-2015 Development Agenda Country Consultations.* http://www.undp.org/content/dam/papua_new_guinea/docs/MDG/UNDP_PG_The%20future%20We%20Want%202015.pdf. Accessed 15 January 2016.

GSMA. 2015. *The Mobile Economy: Pacific Islands 2015.* https://gsmaintelligence.com/research/?file=23485245295f02524925b2bd3aeec6de&download. Accessed 12 Jan 2016.

Howes, S., Mako, A.A., Swan, A., Walton, G., Webster, T. and Wiltshire, C. 2014. *A lost decade? Service delivery and reforms in Papua New Guinea 2002 –2012.* Canberra: The National Research Institute and the Development Policy Centre.

Internet World Stats 2015. Internet Usage and 2015 Population in Oceania: From wealth to wellbeing. http://www.internetworldstats.com/stats6.htm. Accessed 12 Jan 2016.

ITU. 2014. *Measuring the Information Society Report 2014 Highlights.* http://www.itu.int/en/newsroom/Documents/MIS-2014-Highlights.pdf. Accessed 12 Jan 2016.

Kare, P. and Sermel, R. 2013. Providing Relevant and Quality Learning: An Aim of Universal Basic Education. In: Kukari, A. (ed). *Universalizing Basic Education in Papua New Guinea: experiences, lessons learnt, and interventions for achieving the goal of universal basic education*, pp 83-94. Port Moresby: The National Research Institute.

Koczberksi, G. 2007. Loose fruit mamas: Creating incentives for smallholder women in oil palm production in Papua New Guinea. *World Development* 35, 1172-1185.

Koczberski, G., Curry, G.N. and Gibson, K. 2001. *Improving productivity of the smallholder oil palm sector in Papua New Guinea: a socio-economic study of the Hoskins and Popondetta schemes.* Canberra: Department of Human Geography, Research School of Pacific and Asian Studies, Australian National University.

Logan, S. 2012. Rausim! Digital Politics in Papua New Guinea. SSGM Discussion Paper 2012/9. Canberra: Australian National University.

Macintryre, M. 2008. Police and thieves, gunmen and drunks: Problems with men and problems with society in Papua New Guinea. *The Journal of Anthropology* 19(2), 179-193.

Maumbe, B. 2013. Global e-Agriculture and Rural Development: E-Value Creation, Implementation Challenges, and Future Directions. In: Maumbe, B. and Patrikakis, C.Z. (eds), *E-agriculture and Rural Development: Global Innovations and Future Prospects*, pp 1-15. Hershey, Pennsylvania: IGI Global.

McNamara, K., Belden, C., Kelly, T. Pehu, E. and Donovan, K. 2011. Introduction: ICT in Agricultural Development (Module 1). In: World Bank (ed), *ICT in Agriculture Connecting Smallholders to Knowledge, Networks, and Institutions*, pp 3-14. E-sourcebook, The International Bank for Reconstruction and Development/World Bank, Washington. https://www.ictinagriculture.org/content/ict-agriculture-sourcebook. Accessed 10 January 2016.

National Statistical Office (NSO) 2011. Household Income and Expenditure Survey 2009-2010. Summary Tables. Port Moresby: National Statistical Office.

Numbasa, G. and Koczberski, G. 2012. Migration, informal urban settlements and non-market land transactions: a case study of Wewak, East Sepik Province, Papua New Guinea. *Australian Geographer* 43(2), 143-161.

Ogden, M.R. 2013. Communications. In: Rappaport, M. (ed). *The Pacific Islands. Environment and Society*, pp 401-416. Honolulu: University of Hawaii Press

Overfield, D. 1998. An investigation of the household economy: coffee production and gender relations in Papua New Guinea. *Journal of Development Studies* 34(5), 52-72.

PRIF 2015. *Economic and Social Impact of ICT in the Pacific 2015.* Sydney: Pacific Region Infrastructure Facility.

Sharp, T.L.M. 2012. 'Following Buai: the highlands betel nut trade, Papua New Guinea'. Unpubl. PhD thesis, Canberra: Australian National University.

Sinclair, J. 1984. *Uniting a Nation: the Postal and Telecommunication Services of Papua New Guinea.* Melbourne: Oxford University Press.

Storey, D. 2010. *Urban Poverty in Papua New Guinea*, The National Research Institute, Discussion Paper No. 109, Port Moresby.

Suwamaru, J.K. 2013. ICT initiatives in Papua New Guinea: impact of mobile phones on socio-economic development. Unpubl. PhD thesis, Madang: Divine Word University.

Suwamaru, J.K. 2015. Aspects of mobile phone usage in Papua New Guinea: a socio-economic perspective. Contemporary PNG Studies. *DWU Research Journal* 22.

Telban, B., and Vavrova, D. 2014. Ringing the living and the dead: mobile phones in a Sepik society. *The Australian Journal of Anthropology* 3(1), 1-15.

Totora, B. and Rhealt, M. 2011. Mobile Phone Access Varies Widely in Sub-Saharan Africa, Gallup New, September. Available at: http://www.gallup.com/poll/149519/Mobile-Phone-Access-Varies-Widely-Sub-Saharan-Africa.aspx?g_source=position5&g_medium=related&g_campaign=tiles. Accessed 15 January 2016.

UNDP 2013. *Human Development Report 2013: The rise of the south. Human progress in a diverse world.* http://hdr.undp.org/sites/default/files/Country-Profiles/PNG.pdf. Accessed 12 Jan 2016.

UNDP 2014. National Human Development Report. Papua New Guinea. From wealth to wellbeing: Translating resource revenue into sustainable human development. United Nations Development Programme, Port Moresby and Massey University, Auckland.

UNICEF 2013. Statistics at a glance: Papua New Guinea. http://www.unicef.org/infobycountry/papuang_statistics.htm. Accessed 12 Jan 2016.

Waffi, J.M., Atigini, B., Sikas, H. and Hinamunimo, B. 2015. GoPNG – INGO Partnerships: A Case Study of CARE International in Papua New Guinea. http://devpolicy.org/Events/2015/2015-PNG-Update/Presentations/Day-2/Public-private-partnerships_paper_Waffi.pdf. Accessed 15 January 2016.

Wardlow, H. 2006. *Sexuality and Agency in a New Guinea Society: Wayward Women.* University of California Press, Berkeley.

Watson, A.H. 2011. The mobile phone: The new communication drum of Papua New Guinea, West New Britain Province. Unpubl. PhD thesis, Brisbane: Queensland University of Technology.

World Bank 2012a. Papua New Guinea - Country gender assessment for the period 2011-2012. Washington: World Bank.

World Bank 2012b. World Development Report 2012: Gender Inequality and Development. Washington: World Bank.

World Bank 2012c. Information and Communications for Development. Maximising Mobile. Washington: World Bank.

Business, Commerce and the Global Financial System

Meg Elkins and Liam J.A. Lenten

Abstract Commercial practices are being re-defined by disruptive innovations that are opening up new global and local markets. This chapter examines how changing technologies are creating new opportunities for entrepreneurs in both the developing and developed world. In the developing world, micro-finance and mobile technologies are linking the vulnerable to markets. In the developed world long-held monopolies in the banking, transport, and hotel industries are now subject to a more competitive market with the rise of new platforms such as the sharing economy, crypto-currencies, and crowd funding.

Keywords Market access · New financial platforms · Increased competition

Introduction

Individuals and firms previously denied opportunities to participate in the market economy are integrating steadily into the global marketplace through the increased availability of information from the internet and mobile technologies. Micro-finance has given those in the developing world access to credit and micro-enterprises giving the developing world access to the market. In the developed world, the sharing economy, crypto currencies and crowdfunding are disrupting established financial market intermediaries. Internet access and mobile technologies play a significant role in how agents can communicate and interact, particularly in relation to institutions. Intermediaries and asymmetric information are becoming less significant as information and new financial platforms have the capacity to disrupt traditional networks.

M. Elkins (✉)
RMIT University, Melbourne, Australia
e-mail: meg.elkins@rmit.edu.au

L.J.A. Lenten
La Trobe University, Melbourne, Australia
e-mail: l.lenten@latrobe.edu.au

© The Author(s) 2016
M.E. Robertson (ed.), *Communicating, Networking: Interacting*,
SpringerBriefs in Global Understanding, DOI 10.1007/978-3-319-45471-9_6

Micro-Finance—Connecting the Poor to Markets

Micro-finance bridges the gap between the banks and the lenders of money to people in the developing world. These firms have been able to provide small unsecured loans to people who would be ineligible for loans from traditional financial institutions (Yunus 1999). Micro-finance refers to micro-credit, micro-savings, micro-insurance and money transfers for small amounts of money from (USD) $50 to $1000 (Van Rooyen et al. 2012; Yunus 1999). These loans have been attributed to helping micro-entrepreneurs create business and increase income as well as contribute to an improvement in well-being to the life of the poor (Van Rooyen et al. 2012). The Grameen Bank is attributed with the introduction of micro-finance in Bangladesh in the 1970s. It now provides loans valued at $2 billion annually to over 30 million members (Khandker and Samad 2014).

In essence, micro-finance has been able to connect those normally excluded from local and global markets (due to poor asset wealth) to funds many assume essential to get ahead in life. Overwhelmingly, women have benefited the most from gaining access to these funds resulting in greater employment generation, increased income generation and significant improvement in social indicators these indicators include the Human Development Index (HDI), primary school enrolment and health indicators (Khandker 2005; Van Rooyen et al. 2012). Empowering women with businesses to spend their incomes on education and health for their families benefits the individual family most directly, but also has broader ramifications for their society, as it also develops the social and human capital for that society. The establishment of micro-enterprises financed through the micro-credit system creates a more socially-interactive community for the poor to participate in the broader economy.

Mobile Technologies—Connecting Local Businesses to Global Markets

The proliferation of mobile technologies in the developing world has created opportunities for small landholders. Mobile phone networks have provided a more affordable alternative to landlines, which were previously too expensive for villages in remote and regional areas due to the high cost of infrastructure (Dannenberg and Lakes 2013). Farmers with small landholdings have been among the main beneficiaries of the introduction of mobile technologies due to increased access to market information. There are two reasons for this: firstly, village markets are characterised traditionally by asymmetrical information whereby intermediaries or traders are more aware of prices in the central markets. In the past, asymmetrical information has led to low productivity and low farm incomes. In the case of interventions in areas of Kenya, mobile phones have resulted in improvements in food security and farm incomes (Ogutu et al. 2014). Secondly, the introduction of these technologies adjusts the power balance between small-landholder and agricultural intermediaries. Mobile technology allows for immediate access to up-to-date market information, and also provides the farmer with direct access to payments both for paying for agricultural resources, such as fertiliser and seeds; as well as receiving payments for their crops. In rural Kenya, the mobile payment service M-Pesa has been adopted widely by the farmers in the export market for fruit and vegetables (Dannenberg and Lakes 2013). Mobile phones allow farmers to deal directly with exporters, circumventing the need for intermediaries, and ostensibly this gives them access to the global market. Farmers are less likely to accept unfavourable prices for their crops, thus improving their bargaining power (Tadesse and Bahiigwa 2015). Mobile phones also provide important information in regards to best-practice farming, determining best timing of when to plant and when to harvest, as well as market demand for particular crops. All this provides direct benefits to productivity, which should result in higher levels of farm income.

The Changing Role of Markets

There is an increasing recognition of the role and importance of communication, networks and groups in the determination of economic outcomes—employment, income and wealth among these—demonstrative of the shift away from the traditional rational individual (*homo economicus*) assumption that underpins many long-standing models in the discipline.

As both a wonderful reference tool and literature survey, Paul Frijters (Queensland) and Gigi Foster (UNSW) deal with these themes in intimate detail in their recent book (Frijters and Foster 2013) *Economic Theory of Greed, Love, Groups and Networks*. Tellingly, against the background of the basic economic activity of trade, they describe networks as: "…a crucial component of the modern economic system of production and exchange" (p. 5), with Winters et al. (2004) making a useful extension to this. They then proceed to invoke microeconomic theoretical modelling (see pp. 349–396) to extend on several in-text examples.

Frijters and Foster proceeded to cite numerous specific examples of similarly-themed studies in the literature. Some focus on topics that are highly-aligned with identifiable fields within the discipline; such as political economy, public, experimental, development, finance, education, industrial organisation, management, monetary, labour, behavioural, social science and agriculture—see, for instance, Becker (1974) and Caskie (2000) as examples of the latter two. Others are included that cross-over into other (largely exclusive) disciplines, such as psychology, biology, and reproductive science.

The Sharing Economy

The rise of the 'sharing economy' is one of the clearest comparisons between developed and developing nations, with respect to the effect of communication technology on everyday practice. The sharing economy has the potential to disrupt the way we traditionally exchange goods and services and function as markets. This technology is underpinned by the recent growth of application-based programs on mobile communication devices that have decreased the transaction costs of exchange. Specific examples of such applications, and how they have changed the business landscape of selected industries, have already been researched and discussed in the business studies literature; such as the accommodation service AirBnB (Zervas et al. 2015) and the ride-sharing system Uber (Anderson 2014); while more generally, (Belk 2014) reinforces how such innovations force us to re-think the old adage 'you are what you own'. The dynamic pricing model for companies such as Uber provides incentives for those with capital assets (cars) to respond to increases in demand.

In the developed world, such technology plays a significant part in collaborative consumption, essentially between strangers, thus stimulating the volume of everyday practices that typically people would otherwise often undertake in partnership with existing friends, colleagues, associates or family members (ie. within their own network of individuals). Rising incomes in the past few generations have occurred contemporaneously with losses of networks and a sense of community— consistent with other social indicators, such as declining birth rates (Ahn and Mira 2002), smaller household sizes (Australian Bureau of Statistics 2010) and an increasing incidence in mental illness (World Health Organization 2015). Under such trends, these applications have served the potential to fill the void of community breakdown in expanding non-technology collaborative consumption outcomes, which is arguably forecasted to become increasingly necessitated by dwindling global levels of natural resources relative to population.

By stark contrast, in the developing world, with far less (both) human and physical capital, but greater social capital, end users and service providers are typically already connected and have an intimate sense of each other's preferences; while the role of trust is clearly critical. Thus, the traditional levels of collaborative consumption have not reduced significantly during the modern era in the first instance. In this environment, the uptake and impact of these applications is expected to be somewhat more limited, at least in the short-term.

Crypto-Currencies—Financial Markets Without Institutions

One of the next frontiers in global financial systems is that of crypto-currencies. The best known of these currencies is the Bitcoin, which offers alternate methods of exchange without the use of financial institutions. Bitcoin is a system of exchange that is not administered by a single institution, government or country for its existence (Sadeghi 2013). The software underlying its creation established that Bitcoins would be able to be 'mined' slowly and steadily until there was 21 million units in circulation (Cheung et al. 2015). This crypto-currency uses an innovative cryptography called a blockchain; this is a public ledger that is effectively a permanent, incorruptible and irreversible trace of all Bitcoin transactions. In traditional transactions, we use banks to determine all transactions and account balances. The uniqueness of Bitcoin is that an intermediary is no longer required—all individuals consult the most recently-updated public ledger. The Bitcoin revolution is puzzling to economists as intrinsically there is no value of exchange (Yermack 2013). However, its newness and untapped potential could see such transactions be the new protocol of exchange. Bitcoin has the potential to be a revolutionary peer-to-peer platform, akin to the disruption that Skype had on telecommunications; alternatively, this could be merely a passing fad. The biggest barrier to bitcoin and crypto-currencies is the relationship of trust, particularly as these currencies are not managed by a single authorised institution or company.

The Role of the Crowd in the Entrepreneurial Space

Crowdfunding represents another disruptive method by which the internet is offering a new way of engaging the community to support new businesses and ideas. The entrepreneur raises funds by making an open call on the internet collecting small amounts of money from a large number of investors. The crowd, in turn, has a closer relationship with the firm both as consumers and investors (Belleflamme et al. 2014). The equity (profits) of the business is then distributed back to the funding 'crowd'. Kickstarter, Fundable, and Indiegogo are the most well-known of the crowdfunding sites. The pebble 'smart watch' is a case whereby venture capital initially rejected the project but a Kickstarter fund was able to evidence demand for the good and an initial request for US$100,000 for 100 watches resulted in over $US 10 million in pledges (Agrawal et al. 2014). The crowdfunding also serves as a marketing tool to create awareness of products in their development stages (Mollick 2014).

Conclusion

Integration into the marketplace—encouraging entrepreneurial activities in previously untapped markets—allows for global expansion and connectedness. Technological advances are creating new business platforms and financial innovations that are disrupting traditional institutions. Mobile technologies are benefitting consumers and producers in both the developed and developing world. Small landowners are able to make greater profits on their produce by having better access to information in the developing economies. New business platforms are making use of scarce and idle resources such as an individual's home (AirBnB) and car (Uber) with peer-to-peer networks facilitating market transactions in the developed world. The benefit for the consumer is the reduced cost of goods and services. The benefit for the economy is that ordinary individuals are able to leverage businesses off everyday assets, which increases competition among traditional monopolies.

Open Access This chapter is distributed under the terms of the Creative Commons Attribution 4.0 International License (http://creativecommons.org/licenses/by/4.0/), which permits use, duplication, adaptation, distribution and reproduction in any medium or format, as long as you give appropriate credit to the original author(s) and the source, provide a link to the Creative Commons license and indicate if changes were made.

The images or other third party material in this chapter are included in the work's Creative Commons license, unless indicated otherwise in the credit line; if such material is not included in the work's Creative Commons license and the respective action is not permitted by statutory regulation, users will need to obtain permission from the license holder to duplicate, adapt or reproduce the material.

References

Agrawal, A. Catalini, C. and Goldfarb, A. 2014. 'Some Simple Economics of Crowdfunding' in Lerner, J. and Stern, S. (eds.) *Innovation Policy and the Economy, Volume 14*. NBER: University of Chicago Press, 63-97.
Ahn, N. and Mira, P. 2002. 'A Note on the Changing Relationship between Fertility and Female Employment Rates in Developed Countries'. Journal of Population Economics, 15(4), 667-682.
Anderson, D. N. 2014. '"Not Just a Taxi"? For-Profit Ridesharing, Driver Strategies and VMT'. Transportation, 41(5), 1099-1117.
Australian Bureau of Statistics 2010. *2009-10 Year Book Australia*.
Becker, G. 1974. 'A Theory of Social Interactions'. *Journal of Political Economy*, 82(6), 1063-1091.
Belk, R. 2014. 'You are what you can Access: Sharing and Collaborative Consumption Online' *Journal of Business Research*, 67(8), 1595-1600.
Belleflamme, P., Lambert, T. and Schwienbacher, A. 2014. 'Crowdfunding: Tapping the Right Crowd'. *Journal of Business*, 29(5), 585-609.
Caskie, P. 2000. 'Back to Basics: Household Food Production in Russia'. *Journal of Agricultural Economics*, 51(2), 196-209.
Cheung, A., Roca, E. and Su, J.-J. 2015. 'Crypto-currency Bubbles: An Application of the Phillips–Shi–Yu (2013) Methodology on Mt. Gox Bitcoin Prices'. *Applied Economics*, 47(23), 2348-2358.
Dannenberg, P. and Lakes, T. 2013. 'The Use of Mobile Phones by Kenyan Export-orientated Small-scale Farmers: Insights from Fruit and Vegetable Farming in the Mt. Kenya Region'. *Economia Agro-Alementare*, 15(3), 55-76.
Frijters, P. and Foster, J. 2013. *Economic Theory of Greed, Love, Groups and Networks*. Cambridge: Cambridge University Press.
Khandker, S. R. 2005. 'Microfinance and Poverty: Evidence Using Panel Data from Bangladesh'. *World Bank Economic Review*, 19(2), 263-286.
Khandker, S. R. and Samad, H.A. 2014. 'Dynamic Effects of Microcredit in Bangladesh'. *The World Bank, Development Research Group, Working Paper, no. 6821*.

Mollick, E. 2014. 'The Dynamics of Crowdfunding: An Exploratory Study'. *Journal of Business Venturing*, 29(1), 1-16.
Ogutu, S. O., Okello, J. J. and Otieno, D. J. 2014. 'Impact of Information and Communication Technology-Based Market Information Services on Smallholder Farm Input Use and Productivity: The Case of Kenya'. *World Development*, 64, 311-321.
Sadeghi, A. 2013. *Financial Cryptography and Data Security*, 17th International Conference, FC 2013 Okinawa, Japan, April 1-5.
Tadesse, G. and Bahiigwa, G. 2015. 'Mobile Phones and Farmers' Marketing Decisions in Ethiopia'. *World Development*, 68(4), 296-307.
Van Rooyen, C., Stewart, R. and de Wet, T. 2012. The Impact of Microfinance in Sub-Saharan Africa: A Systematic Review of the Evidence. *World Development*, 40(11), 2249-2262.
Winters, L., McCulloch, N. and McKay, A. 2004. 'Trade Liberalization and Poverty: The Evidence so Far'. *Journal of Economic Literature*, 42(1), 72-115.
World Health Organization. 2015. *Mental Health Atlas 2014*.
Yermack, D. 2013. 'Is Bitcoin a Real Currency? An Economic Appraisal'. *NBER Working Paper No. 19747*.
Yunus, M. 1999. *Banker to the Poor: Micro-lending and the Battle against World Poverty*. New York: Public Affairs.
Zervas, G., Proserpio, D. and Byers, J. 2015. *'The Rise of the Sharing Economy: Estimating the Impact of AirBnB on the Hotel Industry'*. Boston University, School of Management, Research Paper No. 2013-16.

Part III
Recommendations—Networking the e-Society

Everyday-ing Health Literacy and the Imperative of Health Communication: A Critical Agenda

Eric Po keung Tsang and Dennis Lai Hang Hui

Abstract Health literacy is an increasingly important issue amongst scholars of health studies and medical practitioners. This essay seeks to understand how health knowledge is co-constructed by different agencies and the role of health communication in this process can contribute to everyday-ing health literacy. In addition, this essay attempts to understand how health communication becomes an everyday practice in identifying disease-specific needs.

Keywords Health literacy · Policy-making · Everyday health needs

Background

To-date, discourses related to health and illness are no longer confined within hospitals and other medical establishments; they have become an everyday, interactive discourse which helps define our way to perceive health issues. This chapter provides an everyday perspective towards the idea of human communication and locates the importance of health literacy—the production and consumption of health-related knowledge and information (Kickbusch and Maag 2008)—in enhancing the quality of public health service. Firstly, we examine the importance of health literacy in establishing the basis of health communication. As recognised by the World Health Organization (n.d.), health literacy shall be playing a critical role in encouraging multi-stakeholder dialogue in empowering individuals and communities. How to put health literacy into policy-based, knowledge-informed action is an important issue that health administrators should consider seriously. Second, we try to look at how health communication is facilitated by the changing

---

E.P.k. Tsang (✉) · D.L.H. Hui
The Education University of Hong Kong, Lo Ping Road, Tai Po, Hong Kong
e-mail: etsang@eduhk.hk

© The Author(s) 2016
M.E. Robertson (ed.), *Communicating, Networking: Interacting*,
SpringerBriefs in Global Understanding, DOI 10.1007/978-3-319-45471-9_7

modes of communication and compare different media-technological options in knowledge construction across different sectors. For the purpose of this chapter, we look specifically at how the increasing concern for health risk offers possibilities for a networked way of knowledge construction with reference to three aspects and the case of mental health literacy. Overall, we attempt to shed light on how information and communications technology can facilitate the promotion of health literacy.

A Conceptual Background About Health Literacy and Health Communication

We begin with the idea of health literacy. Essentially, health literacy refers to the competence to make informed decisions on a wide range of health-related matters such as nutrition, medication, choice of medical services, and ways to minimise exposure to illnesses (Kickbusch and Maag 2008). With the development of information technology and the increasing awareness of the imperatives of public health education, health literacy has become an emerging governance agenda. A new policy discourse has emerged which is driven by participation of different stakeholders such as medical professionals, community networks, research institutions, and individuals. At the normative level, health literacy is about how to create a knowledge discourse which can ensure convergence of values, understandings and expectations about health-related matters. In this connection, World Health Organization points to the need to cultivate 'the cognitive and social skills' in broadening the basis of participation (WHO n.d.). In operationalising the concept, Nutbeam (2000) points to the need to pay attention to functional, interactive and critical aspects of health literacy. Whereas the functional aspect refers to individual's ability to understand basic health information, the interactive aspect refers to the increasing autonomy of individuals to reflect on different health information (ibid). Critical health literacy, meanwhile, transcends those functional parameters and addresses how individuals and communities are networked in health knowledge construction with their own literacies (ibid).

Meanwhile, what has also been explored in the existing literature are the ways in which health literacy can be promoted. Indeed, the unsatisfactory track-record of promoting health literacy at the global level reflects the failure to integrate health, education and communication together. As emphatically argued by Kickbusch (2001), the imbalance in developing competence in health-related matters reflects the inability to overcome the divide between literacy, communication and knowledge construction. In that connection, there is a need to bring health communication back into the policy discourse.

Health communication, broadly speaking, is about how to create a communicative context for achieving a diversity of health-related purposes. As defined by Ratzan et al. (2004), health communication is "the process through which one person, group, or governmental or private organization uses various communication strategies and channels to educate, motivate, and perpetuate information, skills, and behaviours that are generally accepted to benefit (improve) the health of individuals and the public" (p. 398). Originally used in the context of commercial marketing (e.g. the promotion by pharmaceutical companies and insurance companies), health communication has assumed a wider scope of importance (Thomas 2006). Nowadays, health communication becomes an everyday dynamic that defines our understanding about possible options for us to pursue health-related behaviours. For example, health communication is becoming more important in sharing patient information within a jurisdiction (Thomas 2006). Health communication is also widely used in community empowerment (Thomas 2006). In developing economies, for instance, health communications—both formal and informal—have been used to enhance the preparedness towards health crises. In the intellectual context, health communication is shaping the fundamental ways in which health issues are narrated and problematised on a day-to-day basis (Dutta and Zoller 2008). Meanings of health, illness and disease are, accordingly, contingent. A recent edited volume by Hamilton et al. (2014) goes further by contending that health communication is a linguistically situated practice mediated by different contexts and modes of interaction. In this next section, we examine how the advancement of technology has reshaped health communication and the opportunities and challenges associated with promoting health literacy.

Global Diversity in Health Communication and Everyday-ing Health Literacy

Health communication and health literacy co-evolves with the changing form of media (Hagglund et al. 2009). The role of the traditional media in promoting health literacy has been discussed in the existing body of work. Jorm (2000), for example, discusses the use of traditional media in Defeat Depression Campaign in promoting community awareness towards mental health. How far the momentum for health literacy can be sustained is highly debatable. Scholars thus look to the prospect of the ascendency of e-society as a possibility for offering a new way of health communication. For example, emphasising the importance of interactivity and user-friendliness, the new digital media such as the Internet (especially Web 2.0), social media, digital games, and use of avatars has opened up a wider possibility for a networked way for collaborative construction of health information and knowledge (Prestin and Chou 2014). Laverack (2009) argued that with the increasing availability of low-cost and innovative communication options, the role of the public in health literacy is becoming more proactive. Health literacy is, accordingly, more than helping individuals to make informed decision; it becomes a discursive tool for everyday mobilisation for health development. The Centers for Disease Control and Prevention (CDC) has identified the following aspects of social media and the emerging interactive media which allows a more 'credible' form of health communication:

- "Increase the timely dissemination and potential impact of health and safety information.
- Leverage audience networks to facilitate information sharing.
- Expand reach to include broader, more diverse audiences.
- Personalize and reinforce health messages that can be more easily tailored or targeted to particular audiences.
- Facilitate interactive communication, connection and public engagement.
- Empower people to make safer and healthier decisions" (CDC 2011, p. 1).

Based on these aspects, we identify several examples where health communication, health literacy, and the new media coalesce. One of the key observations by public health scholars is that the proliferation of electronic (new) media has led to a 'tectonic shift' in the way people are engaged in processes related to health-related knowledge and information (Hesse et al. 2005). The literature has identified two levels of interaction that the evolving media landscape is impacting on health literacy. On one hand, social media is playing a functional role in expanding the clinical space beyond hospitals and medical establishments. The health care provider is capitalising upon the use of e-technology in forging a closer physician-patient relationship (Sundar et al. 2011). Communications beyond the face-to-face consultation allow patients to receive and share a wider range of clinical information such as chronic disease self-management (Prestin and Chou 2014) and interventional evaluation (Harrison et al. 2010). Meanwhile, it is reported the use of peer-to-peer (P2P) communication is useful in developing competence in understanding technical information and in accessing relevant health provider. P2P communication also provides information related to how patients can communicate with medical specialist. The study by Meier et al. (2007) looks, for example, at how cancer patients can develop linguistic competence in everyday communication with the medical expert. These information are, however, 'instructional' and 'pedagogical' (Evans et al. 2009) at best, and may not promote a more dynamic way of promoting health literacy.

Beyond the clinical dimension, the 'digitalisation' of health communication has created opportunities for a dynamic, critical approach for promoting health literacy. We here offer an overview of three possible opportunities. First is the possible dialogue between 'traditional' expert epidemiology and lay epidemiology. Whereas traditional epidemiology refers to how medical professions define the nature of a health problem based on scientific discourse, lay epidemiology refers to how people construct their health risks with reference to their subjectivities and bodily constitutions. Lay epidemiology is thus about how the society becomes a knowledge basis for possible community-based actions for enhancing health. The possible contribution of public knowledge in informing professional action has been documented in different works (Arksey 1994; Davison et al. 1991). Essentially, it provides very important data about the possible cause of health risks (such as environmental risks). Whilst some scholars have contested the reliability of public knowledge in ensuring effective public awareness, lay epidemiology becomes a new discourse in exposing conceptions (and misconceptions) about health and disease and provides clues as to what kind of public health intervention is necessary (Laverack 2009). This is made possible by the prevailing use of social media of Twitter, Facebook and other social blogs which has fostered a new socio-political culture of sharing.

Another key opportunity is the use of social media in enhancing community preparedness for health disasters. In the first place, the use of geographic information system (GIS) becomes more prevalent in constructing necessary knowledge for emergency medicine. The Ready New York initiative was developed by the New York City Government in allowing residences to develop their own plans of evacuation in cases of natural or human-made disasters (Zarcadoolas and Pleasant 2009). Cromley and McLafferty (2012) point to the geospatial mode of health communication which can allow health authorities, community leaders and residents to identify the possible 'hotspots' for critical intervention. For example, the US Department of Homeland Security's Science & Technology Directorate, capitalising upon the proliferation of smartphones and other GPS technologies, has introduced the Social Media Alert and Response to Threats to Citizens (SMART-C) programme. The programme, accordingly, "aims to develop citizens' participatory sensing capabilities for decision support throughout the disaster life cycle via a multitude of devices (such as smartphones) and modalities (MMS messages, Web portals, blogs, tweets, and so on)" (Adam et al. 2012, p. 92). At the same time, the use of online virtual gaming is popularising emergency health knowledge. The Centers for Disease Control and Prevention has recently made use of Zombie Preparedness to create a virtual community for engaging public in developing emergency preparedness (CDC 2015).

The third opportunity is the usage of new media in different health campaigns can equip community members for enhancing linguistic and knowledge capacities. Although the traditional printed media has a relatively long history in contributing to the success of many health campaigns, the new media provides a more dynamic and robust way of engaging a wider scope of audience. Yan Tian's study on the impact of Web 2.0 arrives at the conclusion that the media has a positive impact on the audience's understanding about issues related to organ donation (Tian 2010). Donors' families, recipients and activists are able to make use of the digital media platform such as YouTube for disseminating the medical knowledge and benefits of organ donation (ibid). Meanwhile, new media has been argued to have played a crucial role in the promotion of sexual health (Guse et al. 2012). The ability to identify the health risks associated with unsafe sexual practices has accordingly been enhanced with the digitalisation of health communication (ibid). On the whole, all these three opportunities have pointed to the fact that the role of the new media in promoting public health goes beyond passive acquisition of health-related information; instead, it points to the possibility for a decentred, non-hegemonic way of disseminating and expanding the existing body of health knowledge.

Case Study: Social Media and the Promotion of Mental Health Literacy

Whilst mental health has been a growing concern amongst many communities, public health knowledge about this issue has been unsystematic and poor in terms of quality (Jorm 2000; Reavley and Jorm 2013). As lamented by Jorm (2000), "members of the public cannot correctly recognise mental disorders and do not understand the meanings of psychiatric terms" (p. 396). This gives rise to the problem of miscommunication between the public and the medical professionals (ibid). Against this context, there has been a call for paying attention to mental health literacy. As argued by Reavley and Jorm (2013, p. 51), "although mental health literacy incorporates knowledge about mental disorders, it goes further in that it places emphasis on the knowledge being linked to the possibility of action to benefit one's own mental health or that of others." Mental health literacy should thus focus on "(1) knowing how to prevent mental disorders; (2) recognition of when a disorder is developing to facilitate early help-seeking; (3) knowledge of help-seeking options and available treatments; (4) knowing effective self-help strategies for milder problems; (5) first aid skills to support others who are developing a mental disorder or in a mental health crisis" (ibid, p. 51).

Meanwhile, how to operationalise the idea of mental health literacy is still a contested issue. One of the key opportunities since the 1990s is the changing mode of communication which renders mental health no longer a 'taboo' to discuss. There have even been more initiatives taken by different communities to address the need to promote mental health literacy. Initiatives such as Mental Health Foundation and SANE Australia, for example, are well-established organisations in promoting mental health knowledge. SANE Australia, for example, has established a 'Lived Experience Forum' which provides 'a safe, anonymous place' for the public to share their concern (SANE Australia, n.d., a). Both Mental Health Foundation and SANE Australia have made full use of new media to disseminate knowledge about keeping mental healthiness. For example, the former organisation has designed free podcasts which introduce a variety of lifestyle tips for the public (Mental Health Organisation, n.d.). SANE Australia, on the other hand, has developed a guide to the media in order to avoid any further stigmatisation on those who are suffering from mental illness (SANE Australia, n.d., b).

Other initiatives take a more critical stance about the way in which mental health knowledge and information is constructed. Formed by a group of health activists, Hearing Voices Network (2015) has pointed to the 'unsoundness' and 'unreliability' of the contemporary approach of psychiatric diagnoses: "We believe that people with lived experience of diagnosis must be at the heart of any discussions about alternatives to the current system. People who use services are the true experts on how those services could be developed and delivered; they are the ones that know exactly what they need, what works well and what improvements need to be made" (Hearing Voices Network 2015). The Network organises an online platform on which those who are suffering from emotional disturbances can express themselves without emotional burdens. As asserted by Pilgrim and Rogers (2003), whilst initiatives of this kind may not be scientifically measurable, they do provide a possibility for 'emancipation' and for challenging the dominance of professional knowledge in understanding the medical condition of those who suffer from similar symptoms.

Open Access This chapter is distributed under the terms of the Creative Commons Attribution 4.0 International License (http://creativecommons.org/licenses/by/4.0/), which permits use, duplication, adaptation, distribution and reproduction in any medium or format, as long as you give appropriate credit to the original author(s) and the source, provide a link to the Creative Commons license and indicate if changes were made.

The images or other third party material in this chapter are included in the work's Creative Commons license, unless indicated otherwise in the credit line; if such material is not included in the work's Creative Commons license and the respective action is not permitted by statutory regulation, users will need to obtain permission from the license holder to duplicate, adapt or reproduce the material.

References

Adam, N.R., Shafiq, B. and Staffin, R., 2012. Spatial Computing and Social Media in the Context of Disaster Management, *Intelligent Systems, IEEE*, 27(6), 90-96.
Arksey, H., 1994. Expert and lay participation in the construction of medical knowledge, *Sociology of Health and Illness*, 13(1) 1-19.
Centers for Disease Control and Prevention (CDC), 2011, *The Health Communicator's Social Media Toolkit*, available at: http://www.cdc.gov/healthcommunication/ToolsTemplates/SocialMediaToolkit_BM.pdf (accessed 12 August 2015).
Centers for Disease Control and Prevention (CDC), 2015. "Zombie preparedness," available at: http://www.cdc.gov/phpr/zombies.htm (accessed 16 August 2015).
Cromley, E. and McLafferty, S., 2012. *GIS and Public Health*, New York: The Guilford Press.
Davison, C., Smith, G., Frankel, S., 1991. Lay epidemiology and the prevention paradox: the implications of coronary candidacy for health education, *Sociology of Health and Illness*, 16(4), 448-468.
Dutta, M. and Zoller, H., 2008, Theoretical Foundations: interpretative, critical, and cultural approaches to health communication. In H. Zoller and M. J. Dutta, eds., *Emerging Perspectives in Health Communication: Meaning, Culture, and Power*, pp. 1-28. Oxon: Routledge.
Evans, J., Davies, B. and Rich, E., 2009. The body made flesh: Embodied learning and the corporeal device, *British Journal of Sociology of Education*, 30(4), 391-505.
Guse, K., Levine, D., Martins, S., Lira, A., Gaarde, J., Westmorland, W. and Gilliam, M., 2012. Interventions Using New Digital Media to Improve Adolescent Sexual Health: A Systematic Review, *Journal of Adolescent Health*, 51(6), 535-543.
Hagglund, K., Shigaki, C., and McCall, J. 2009. New Media: A Third Force in Health Care. In Jerry C. Parker, Esther Thorson (eds.), *Health Communication in the New Media Landscape*, pp. 417-36. New York : Springer.
Hamilton, H., and Chou, Wen-ying Sylvia eds. 2014. *The Routledge Handbook of Language and Health Communication*, Oxon: Routledge.
Harrison, T., Morgan S., King, A., et al., 2010. Promoting the Michigan Organ Donor Registry: Evaluating the Impact of a Multifaceted Intervention Utilizing Media Priming and Communication Design, *Health Communication*, 25(8), 700-708.
Hearing Voices Network 2015. Position Statement on DSM 5 & Psychiatric Diagnosis, available at: http://www.hearing-voices.org/about-us/position-statement-on-dsm-5/ (accessed 17 August 2015).

Hesse, B., Nelson, D., Kreps, G. et al., 2005. Trust and Sources of Health Information: The Impact of the Internet and Its Implications for Health Care Providers: Findings From the First Health Information National Trends Survey. *Archives of Internal Medicine*, 165(22), 2618-2624.

Jorm, A. 2000. Mental health literacy: public knowledge and beliefs about mental disorders, *British Journal of Psychiatry*, 177, 396-401.

Kickbusch, I. 2001. Health literacy: addressing the health and education divide, *Health Promotion International*, 16(3), 289-97.

Kickbusch, I. and Maag, D., 2008. Health Literacy. In K. Heggenhougen, S. Quah, eds., *International encyclopaedia of public health*, pp. 204–211. Boston : Elsevier.

Laverack, G. 2009. *Public Health: Power, Empowerment and Professional Practice*, Basingstoke, Hampshire: Palgrave Macmillan.

Meier A, Lyons E, Frydman G, Forlenza M and Rimer B, 2007. How Cancer Survivors Provide Support on Cancer-Related Internet Mailing Lists, *Journal of Medical Internet Research*, 9(2), e12.

Mental Health Organisation, (n.d.), Mental Health Organisation, available at: http://www.mentalhealth.org.uk/help-information/podcasts/ (accessed 12 August 2015).

Nutbeam, D., 2000. Health literacy as a public health goal: A challenge for contemporary health education and communication strategies into the 21st century, *Health Promotion International*, 15(3), 259-267.

Nutbeam, D., 2008. The evolving concept of health literacy," *Social Science and Medicine*, 67(12) 2072-2078.

Parrott, R., Volkman, J. E., Lengerich, E., Ghetian, C. B., Chadwick, A. E. and Hopfer, S. 2010. Using Geographic Information Systems to Promote Community Involvement in Comprehensive Cancer Control, *Health Communication*, 25(3), 276-285.

Pilgrim, D., and Rogers, A., 2003, *Mental Health and Inequality*, Basingstoke, Hampshire: Palgrave Macmillan.

Prestin, A., and Chou, W-Y S. 2014, Web 2.0 and the changing health communication environment. In Heidi Hamilton, Wen-ying Sylvia Chou (eds.), *The Routledge Handbook of Language and Health Communication*, pp. 184-197. Oxon: Routledge,

Ratzan, S., Payne, J., and Schulte, S. 2004. Health communication. In N. Anderson (Ed.), *Encyclopedia of Health and Behavior*, pp. 398-402. Thousand Oaks, CA: SAGE Publications, Inc.

Reavley, N., and Jorm, A., 2013. Mental health literacy. In L. Knifton, and N. Quinn. *Public Mental Health: Global Perspectives*, pp. 50-58. Maidenhead, Open University Press.

SANE Australia, (n.d., a), "Lived Experience Forum," available at: http://saneforums.org/t5/Lived-Experience-Forum/ct-p/lived-experience-forum (accessed 16 August 2015).

SANE Australia, (n.d.,b), "How media professionals can reduce the impact of prejudice and discrimination on people affected by mental illness," available at: https://www.sane.org/images/media-centre/brochure_stigma_media.pdf (accessed 16 August 2015).

Sundar, S., Rice, R., Kim, H-S. and Sciamanna, C., 2011. Online Health Information. In T. Thompson, R. Parrott, and J. Nussbaum (eds.), *Routledge Handbook of Health Communication* (2nd edition), pp. 181 202. Oxon: Routledge.

Thomas, R., 2006. *Health Communication*, New York: Springer.

Tian, Y. 2010. Organ Donation on Web 2.0: Content and Audience Analysis of Organ Donation Videos on YouTube. *Health Communication*, 25(3), 238-246. doi:10.1080/10410231003698911

World Health Organization, (n.d.), "Track 2: Health literacy and health behaviour," available at: http://www.who.int/healthpromotion/conferences/7gchp/track2/en/ (accessed 16 August 2015).

Zarcadoolas, C. and Pleasant, A., 2009. Health Literacy in the Digital World. In J. C. Parker, E. Thorson (eds.), *Health Communication in the New Media Landscape*, pp. 303-323. New York: Springer.

Imaging an E-future: Education as a Process Towards Understanding

Margaret E. Robertson

Abstract Education is the vital tool for understanding the complexities of living within social structures—regardless of scale. Traditions that shaped the infrastructure of ancient communities leave a residue presence today in architectural, cultural, economic and political values and practices. The past helps us comprehend the rapid changes to the space, place and environment interactions associated with transport and communications developments. Nowadays time-space compression is reorganising our global networks and commodity flows. Understanding events in the twenty-first century requires communities to adjust but preserve their collective memory.

Keywords Ways of knowing · Education · Understanding · Ecological change · Memory preservation

Towards Understanding

New geographies are products of time and space reorganisation. The history of human occupancy on earth highlights periods of great change caused by migration, trade and colonisation of lands distant from 'home'. These periods facilitated the spread of ideas and cultural, economic and political practices. At the heart of much of the interaction has been trade. Hence, colonisation, imperialism (mostly of European origins), capitalism associated with the free market economy, and the production of monetary wealth, are key processes for understanding global transactions. The argument we make is that whilst new technologies enable transactions by anyone, anywhere, anyplace and across traditional boundaries, the processes that

---

M.E. Robertson (✉)
La Trobe University, Melbourne, Australia
e-mail: m.robertson@latrobe.edu.au

© The Author(s) 2016
M.E. Robertson (ed.), *Communicating, Networking: Interacting*,
SpringerBriefs in Global Understanding, DOI 10.1007/978-3-319-45471-9_8

underpin each transaction—small and large—are similar. However, this assumes the universality of the European and United States superpower discourse (Said 1994). In Said's words:

> Without significant exception the universalizing discourses of modern Europe and the United States assume the silence, willing or otherwise, of the non-European world. There is incorporation, there is inclusion; there is direct rule; there is coercion. But there is only infrequently an acknowledgement that the colonised people should be heard from, their ideas known. (Said 1994, p. 50)

Understanding processes that lead to 'native nationalism' (Said 1994, p. 51), will undoubtedly, help explain contemporary geo-political discourses. And, whilst emergence from colonial rule by modern nation-states is a field of scholarship beyond the scope of this book we do acknowledge the powers of contemporary communications process helping to shape local communities and give voice to their cultural needs. In this book, case studies from Sierra Leone and Papua-New Guinea provide illustrations of innovative practices bringing about change and improving lives within local communities. Worthy of celebration for showing how new technologies can help transform local health, education and daily survival, they also highlight the dangers of assumptions reliant on centralised managerialism, without adequate consultation and capacity building at local levels. Two-way education helps to maintain traditions and cultural identity within a framework that can benefit from modern infrastructure and preserve the 'landscape memory' (see Schama 1995; Lowenthal 2015).

Whilst this may be good advice, finding 'voice' to negotiate better futures remains a struggle for many indigenous societies. Help is not necessarily derived from central agencies. However, being connected to the worlds of others via the internet can provide agency to expand upon traditional knowledge, and ideally, evolve into contemporary lifestyles in harmony with neighbours—near and far. Regrettably, the time-space compression affordances of new technologies are not all for the good of humanity. Internet tools can, and are, fuelling contemporary crises of faith. The age appears to have encouraged defection from mainstream values and the privileging of subjective politics where 'wrong' can appear to outweigh social good. In this sense there is a politicization of the ideals of democracy for personal gain (Žižek 1999). As Keen (2015) asserts in his book titled *The Internet is not the answer* the surveillance capacity of big data services such as *Facebook* and *Instagram* are "creating a panopticon of information gathering" (p. xiv) which: "Rather than focussing in renaissance, it has created a self-centered culture of voyeurism and narcissism" (ibid.).

Why? Arguably, the pace of information flows made possible with new technologies has overwhelmed the political processes. The conditions for knowledge transfer have changed and in turn systems have to change to accommodate new spatial and temporal realities. Transdiscipline approaches are needed, and perhaps, as David Harvey concedes in his most recent book (2016) we need to consider a 'critical anti-capitalist' world. Neo-liberalism and capitalist agendas may be both the high and low points of the Anthropocene.

Along with the continual making of new geographies, including the concretization of the landscape as human populations increasingly live in mega-cities, the draw down on survival physical resources is also threatening the environment. Nature and culture intertwined are elemental concepts for human survival. Climate change is real. Warming of our oceans and air currents is directly linked to the symbols of modern living—car ownership, long-distance travel, larger homes containing multiple energy using appliances, and perhaps of greatest impact for families—access to lines of monetary credit. Critical theorists and scholars like Harvey (2016) are now raising the alarm for agencies to consider almost the unthinkable for future survival. A right hand turn from capitalism may help save the planet but any slowing of the market driven juggernaut will require education of the current millennial generation. What follows is a summary analysis of some practices which use new learning tools whilst simultaneously encouraging responsible lifestyle choices.

Everyday Knowing and Education in the Millennial Age

Migration shifts to big cities and concentrations of people into even bigger cities is reshaping our urban traditions and further depleting the rural landscapes of labour. Farm workers and their traditions are being reshaped by technologies with labour force reduction being a significant outcome. Whilst labour force depletion in agricultural pursuits is set to continue[1] the need for workers to build infrastructure in new and expanding cities—most notably in East Asia—offers diverse opportunities for alternative employment. For instance, in its Urban Development Series, the World Bank Group (2015) surveys of urban 'spatial growth' in East Asia shows how big data, geospatial technologies and visual imagery are helping the urbanisation transition process. This intersection of technology helps to improve everyday lives in rural areas and contributes to better outcomes in the construction and reconstruction of urban spaces. Knowledge dissemination in the technically connected age also helps transfer skills on how to make improvements that reduce carbon footprint and foster new ecologies for living based on principles of sustainability. Urban gardens, and learning how to garden, are skills for everyday living which can be learnt through community based programs and action based learning in schools. Rather than being lost from memory and identity smart technologies can help younger generations adapt the skills of their rural forebears into current food production lifestyles suitable for urban survival.

Signs of what is shaping up to be a self-help way of living are already operating in the global economy. Uber, AirBnB, online personal and food shopping, are global systems changing the ways in which people interact with one another and

---

[1]See The World Bank prediction that 'Today's urban population of about 3.5 billion people is projected to reach 5 billion by 2030, with two-thirds of the global population living in cities.' Retrieved June, 2016 from http://www.worldbank.org/en/topic/urbandevelopment.

their respective communities. These are all signs on an increasing independence within populations and willingness to reach beyond the mainstream systems. Entrepreneurial in their respective ways each system is creating new economies and shifting the balance of consumer interactions. This can be disruptive for business being conducted in traditional ways and already there as signs of regulatory frameworks appearing from governments reacting to community lobby groups including local taxi companies and hotel chains. In the contemporary age of mobility and growth across real space hegemonic power displacement is a likely casualty. Personal geographies operating in juxtaposition with traditional values and day to day operating systems can be viewed as fracturing the orderly conduct of community life. Uncertainty in government or slow responses to community driven disruption can have negative and positive implications. Where personal or group initiatives go against the accepted societal norms the collective memory of the people can be threatened. Optimistic thinking offers an alternative. Innovation and creative use of the tools of the age can transform lives and help tackle poverty, abuse and unemployment. The imperative for education is to encourage and support leaders who are charismatic in style but guided by humanitarian principles and social justice outcomes.

Final Recommendations

The passage of time inevitability shapes place and space connections. They change constantly. Just as history helps us to understand the dynamics of people and place interactions collective memory is useful but not an exhaustive bowl of wisdom. Different paradigms or discourses operate simultaneously and understanding the processes at work within localised communities is complex. Social dissent, hybrid cultures, mobility and environmental stress are all sources of community upheaval. In the current anthropogenic period, with the right decisions, any of these dynamics can precipitate an evolution in lifestyles to harmonious outcomes between nature and culture (Foucault 1994; Whitehead 1971). The alternatives of disunity and chaos are also possible. The pace of change and the opportunities to connect people and place through digital networks is an energy source like no other in history. Education imaginaries that reflect the digital age are the major defence humanity has for better futures. At the same time structures are needed for daily life to operate efficiently. No matter the scale—global, national, regional and local—shared value systems are imperatives for education.

In the International Year of Global Understanding the challenge for education is to capture the views of the majority perspectives and be inclusive of minority voices. The struggle for global understanding will require monumental effort from within and outside mainstream organisations including governments, corporate and small scale enterprises. This is not simply a matter for the millennial generation alone.

Open Access This chapter is distributed under the terms of the Creative Commons Attribution 4.0 International License (http://creativecommons.org/licenses/by/4.0/), which permits use, duplication, adaptation, distribution and reproduction in any medium or format, as long as you give appropriate credit to the original author(s) and the source, provide a link to the Creative Commons license and indicate if changes were made.

The images or other third party material in this chapter are included in the work's Creative Commons license, unless indicated otherwise in the credit line; if such material is not included in the work's Creative Commons license and the respective action is not permitted by statutory regulation, users will need to obtain permission from the license holder to duplicate, adapt or reproduce the material.

References

Foucault, M. 1994. *The order of things. An anthropology of the human sciences*. New York: Vintage Books.
Harvey, D. 2016. *The ways of the world*. London: Profile Books.
Keen, A. 2015. *The internet is not the answer*. London: Atlantic Books.
Lowenthal, D. 2015. *The past is a foreign country – revisited*. Cambridge: Cambridge University Press.
Said, E. W. 1994. *Culture and imperialism*. New York: Vintage Books.
Schama, S. 1995. *Landscape and memory*. New York: Vintage Books.
Whitehead, A.N. 1971. *Concept of nature*. London: Cambridge University Press.
World Bank Group. 2015. *East Asia's Changing Urban Landscape : Measuring a Decade of Spatial Growth*. Urban Development; Washington, DC: World Bank. Online at https://openknowledge.worldbank.org/handle/10986/21159 License: CC BY 3.0 IGO."vvv
Žižek, S. 1999. *The ticklisk subject. The absent centre of political ontology*. London: Verso.

Epilogue

Margaret E. Robertson

> *The art of progress is to preserve order amid change and to preserve change amid order.*
>
> Alfred North Whitehead (1861–1947)

Abstract Interacting, networking and communicating are essential components of everyday life. Finding solutions for peaceful co-existence between all the world's peoples is fundamental for better lives and the preservation of planet earth.

Keywords Change · Innovation · Better futures

Interacting, communicating and networking are key dimensions for enhancing global understanding. The histories of the world's peoples are a rich source of wisdom for how to achieve contentment in everyday life. Customs passed on through the generations give purpose to our experiences and guide us towards wise decisions. However, whilst the traditions and values of our forebears are a vital source of learning and knowledge, the past alone is seldom enough to satisfy human appetites. Curiosity about the world around us is a constant source of wonder, creativity and discovery. New knowledge adds to the available tools at hand to solve new problems and improve lives. Whitehead's words on progress and change capture the dynamic of existence and serve as a reminder of the duality faced during the change process. Change during this millennial period may have disrupted many of the old order customs – especially in relation to communication and mobility options – but the future is ours to create.

Margaret E. Robertson (Editor, June 2016)

Open Access This chapter is distributed under the terms of the Creative Commons Attribution 4.0 International License (http://creativecommons.org/licenses/by/4.0/), which permits use, duplication, adaptation, distribution and reproduction in any medium or format, as long as you give appropriate credit to the original author(s) and the source, provide a link to the Creative Commons license and indicate if changes were made.

The images or other third party material in this chapter are included in the work's Creative Commons license, unless indicated otherwise in the credit line; if such material is not included in the work's Creative Commons license and the respective action is not permitted by statutory regulation, users will need to obtain permission from the license holder to duplicate, adapt or reproduce the material.

M.E. Robertson (✉)
Department of Education, La Trobe University, Bundoora, VIC, Australia
e-mail: m.robertson@latrobe.edu.au

The manufacturer's authorised representative in the EU is Springer Nature Customer Service Centre GmbH, Europaplatz 3, 69115 Heidelberg, Germany. If you have any concerns regarding our products, please contact ProductSafety@springernature.com

Printed and bound by CPI Group (UK) Ltd, Croydon, CR0 4YY

23/03/2026

02076360-0006